# OXFORD CHEMISTRY MASTERS

## Series Editors

RICHARD G. COMPTON
University of Oxford

STEPHEN G. DAVIES
University of Oxford

JOHN EVANS
University of Southampton

# OXFORD CHEMISTRY MASTERS

# Combinatorial Chemistry

**NICHOLAS K. TERRETT**
*Pfizer Central Research, Sandwich*

Oxford   New York   Tokyo
OXFORD UNIVERSITY PRESS
1998

*Oxford University Press, Great Clarendon Street, Oxford OX2 6DP*

*Oxford   New York*
*Athens   Auckland   Bangkok   Bogota   Bombay*
*Buenos Aires   Calcutta   Cape Town   Dar es Salaam*
*Delhi   Florence   Hong Kong   Istanbul   Karachi*
*Kuala Lumpur   Madras   Madrid   Melbourne*
*Mexico City   Nairobi   Paris   Singapore*
*Taipei   Tokyo   Toronto   Warsaw*

*and associated companies in*
*Berlin   Ibadan*

*Oxford is a trade mark of Oxford University Press*

*Published in the United States*
*by Oxford University Press Inc., New York*

*A catalogue record for this book is available from the British Library*

*Library of Congress Cataloging in Publication Data*
*(Data applied for)*

*ISBN 0 19 850220 6*

*Typeset by EXPO Holdings, Malaysia*

*Printed in Great Britain by*
*Bath Press Ltd., Bath, Avon*

# Series Editors' Foreword

The field of combinatorial chemistry was instigated primarily in response to the need for large libraries of novel compounds to feed the high throughput screening facilities developed by biologists. Traditional methods of organic synthesis involving the sequential synthesis of individual molecules are simply orders of magnitude too slow to satisfy the demand for compounds.

Combinatorial chemistry is still in its infancy, developing from the synthesis of large mixtures of compounds to the high speed parallel synthesis of large libraries of well characterized single compounds, with at the same time a shift in emphasis from solution phase to solid phase methodologies.

Synthetic chemistry is being revolutionized. The development of the shortest most efficient route to particular classes of compounds is no longer *de rigueur*: the skill is now to devise syntheses that maximize the potential for introducing diversity into the product libraries. Polymer chemistry has been revitalized by the realization that in many cases it is the substrate rather than the reagent that needs to be immobilized. The portfolio of new linkers and efficient reactions that can be achieved in solid phase chemistry is rapidly increasing, allowing longer and longer syntheses to be achieved and ever more sophisticated libraries to be produced. New processes allowing the automation of synthesis and analysis are making the libraries ever more valuable. The concomitant need for new information management systems to handle the vast amounts of data is also being addressed.

Oxford Chemistry Masters are designed to provide clear and concise accounts of established and emergent important topics that may be encountered by research chemists at any stage of their career. They are designed to provide an overview of the basic principles and the state of the art in the topic. Combinatorial chemistry is here to stay and this contribution to the Oxford Chemistry Masters by a very significant contributor to its development, Nick Terrett, is sure to remain an essential text to new entrants to the area as well as to old hands.

Stephen G. Davies
Richard G. Compton
John Evans

# Preface

Combinatorial chemistry and the synthesis of compound libraries encompass a broad range of highly varied chemical techniques with an exceptionally wide scope of applications. The growth of interest in the differing expressions of this technology stems from the realization that this, beyond many other chemical techniques, actually provides significant practical and economic advantages. Although the technology is currently used in both academic and industrial contexts, the impact on the pharmaceutical industry especially, has developed 'combichem' from a laboratory curiosity to a key enhancement in the process of drug discovery. So widespread is the impact of this technology that it is now commonly found as part of university undergraduate degree courses. Every pharmaceutical research laboratory has at least one team working in the area, and a host of specialist biotechnology companies have sprung up to exploit the technology.

Combinatorial chemistry has initiated a major rethink of the way chemistry is practised. Several long held assumptions have been challenged and many rules have been rewritten. In particular, the field has initiated a reappraisal and consequent renaissance in the synthesis and analysis of compounds attached to solid phase supports. These aspects are irrevocably changing the way chemistry is performed, and will lead ultimately to more rapid and effective research.

This book aims to provide a basic introduction to the field of combinatorial chemistry, describing the development of major techniques and applications. In addition, as is common to any rapidly developing technological field, there is a requirement to define the many neologisms that have been coined to describe the methods. The book will explain why some techniques are now commonly used and others remain for experts only. Many types of combinatorial chemistry require specialist equipment or skills to use effectively, but their description has been included to provide both a historical perspective as well as to illustrate the considerable diversity in the range of philosophies and technologies that combinatorial chemistry represents. The challenges presented by high speed chemistry have been tackled by an extraordinarily innovative use of multi-disciplinary approaches and in this respect its study reflects the current trend for chemists to become knowledgeable practitioners of several overlapping scientific fields. In the life sciences at least, chemistry in isolation is no longer a satisfactory science, and it is in the wide and varied applications of synthesis that the greatest scientific progress is currently being made.

The age of high speed chemistry is here, and solid phase techniques and parallel synthesis are readily accessible to any skilled chemist with minimal new equipment. Of the many applications of combinatorial chemistry, it has been in drug discovery that the farthest reaching effects of this technology are being felt. Consequently, there is a particular focus on the use of combinatorial chemistry techniques in drug discovery, although as will be described, the development of compound library techniques owe much to researchers in immunology, and recent library applications have been in fields as diverse as molecular recognition, medical diagnostics, chemical process optimization, and even the chemistry of semiconductors. All these areas are touched upon. The only major omission, and as a book with a focus on synthetic chemistry, this is necessarily deliberate, is the use of bacteriophage display libraries, a topic that is comprehensively covered in books and reviews elsewhere, and although of great academic and practical interest, is not a subject of general relevance to practising organic synthetic chemists.

The fundamentals of combinatorial chemistry apply irrespective of the particular field of use, and it is hoped that the book will thus serve as a useful guide for chemists wishing to gain a grounding in this area, as well as a reference source for those already engaged in the exploitation of combinatorial libraries. Because this is a huge field that crosses several disciplines, it has not been possible to describe every application in detail. Consequently, I would recommend that the interested reader consult the references for further background. The primary literature is the main source of detailed experimental information as well as the philosophy of combinatorial chemistry, and I have attempted to make the references as detailed and comprehensive as possible.

Finally, I should acknowledge the ideas and support of Tony Baxter who was involved in this book from the start, and my knowledgeable colleagues Mark Ashton, Gurdip Bhalay, Andy Boyd, Tom Coulter, Martin Edwards, Neal Hone, and Martin Scobie for their constructive comments.

*Sandwich*                                                                    N.K.T.
January 1998

# Contents

# Abbreviations

| | |
|---|---|
| 7TM | seven transmembrane (receptor) |
| ACE | angiotensin converting enzyme |
| ACE-Cl | $\alpha$-chloroethyl chloroformate |
| ACE MS | affinity capillary electrophoresis mass spectrometry |
| ADCC | 4-acetyl-3,5-dioxo-1-methylcyclohexane carboxylic acid |
| AIBN | 2,2'-azobisisobutyronitrile |
| ATPase | adenosine triphosphatase |
| ATR | attenuated total reflectance |
| BAPA | $N$-$\alpha$-benzoyl-D,L-arginine-$p$-nitroanilide |
| Boc | *tert*-butyloxycarbonyl |
| cAMP | cyclic adenosine monophosphate |
| CAN | ceric ammonium nitrate |
| CE | capillary electrophoresis |
| COSY | correlation spectroscopy |
| CPG | controlled pore glass |
| DBU | 1,8-diazobicyclo[5.4.0]undec-7-ene |
| DCC | 1,3-dicyclohexylcarbodiimide |
| Ddz | $\alpha,\alpha$-dimethyl-3,5-dimethoxybenzyloxycarbonyl |
| DEAD | diethyl azodicarboxylate |
| DIC | 1,3-diisopropylcarbodiimide |
| DIPEA | diisopropylethylamine |
| DMAP | dimethylaminopyridine |
| DMF | dimethylformamide |
| DMPU | 1,3-dimethyl-3,4,5,6-tetrahydro-2(1H)-pyrimidinone |
| DMSO | dimethylsulfoxide |
| DMT | 4,4'-dimethoxytrityl |
| DNA | deoxyribonucleic acid |
| DRIFTS | diffuse reflectance infrared Fourier transform spectroscopy |
| EEPROM | electrically erasable, programmable read-only memory |
| ELISA | enzyme-linked immunosorbant assay |
| ESI | electrospray ionization |
| FAB | fast atom bombardment |
| FACS | fluorescence-activated cell sorting |
| FKBP | FK-506 binding protein |
| FMDV | foot and mouth disease virus |
| Fmoc | fluorenylmethoxycarbonyl |
| FT–IR | Fourier transform infrared (spectroscopy) |
| FTICR | Fourier transform ion cyclotron resonance |
| GA | genetic algorithm |
| GC | gas chromatography |
| GC–MS | gas chromatography—mass spectrometry |
| GPCR | G-protein coupled receptor |
| h-EGF | human epidermal growth factor |
| HA1 | haemagglutinin protein |
| HBTU | 2-(benzotriazol-1-yl)-1,1,3,3-tetramethyluronium hexafluorophosphate |

| | |
|---|---|
| HEMA | 2-hydroxyethyl methacrylate |
| HF | hydrogen fluoride |
| HIV-1 | human immunodeficiency virus-1 |
| HMP | hydroxymethylphenoxy |
| HMPA | hydroxymethylphenoxyacetic acid |
| HOBt | *N*-hydroxybenzotriazole |
| HPLC | high performance liquid chromatography |
| HRV | human rhinovirus |
| HSV-1 | herpes simplex virus-1 |
| $IC_{50}$ | inhibition constant (concentration that gives 50% inhibition) |
| ICE | interleukin converting enzyme |
| IR | infrared (spectroscopy) |
| $K_d$ | dissociation constant |
| $K_i$ | inhibition constant |
| LC–MS | liquid chromatography—mass spectrometry |
| LOSC | laser optical synthesis chip |
| mAb | monoclonal antibody |
| MALDI–TOF | matrix assisted laser/desorption ionization—time-of-flight |
| MAS | magic angle spinning |
| MBHA | 4-methylbenzhydrylamine |
| mCPBA | *meta*-chloroperbenzoic acid |
| MeO–PEG | polyethylene glycol methyl ether |
| MS | mass spectrometry |
| MTBD | 7-methyl-1,5,7-triazabicyclo[4.4.0]dec-5-ene |
| $NK_2$ | neurokinin-2 |
| $NK_3$ | neurokinin-3 |
| NMMO | *N*-methylmorpholine oxide |
| NMR | nuclear magnetic resonance |
| NPIT | nitrophenylisothiocyanate-*O*-trityl |
| NSG | *N*-substituted glycine |
| NVOC | nitroveratryloxycarbonyl |
| PCR | polymerase chain reaction |
| PEG | polyethylene glycol |
| PKA | protein kinase A |
| $pK_a$ | —log(association constant) |
| PTP1B | protein tyrosine phosphatase 1B |
| REM | <u>r</u>egenerable resin linker initially functionalized via a <u>M</u>ichael reaction |
| RF | radiofrequency |
| RNA | ribonucleic acid |
| RT | reverse transcriptase |
| SAR | structure–activity relationship |
| SASRIN | super acid-sensitive resin |
| SCAL | safety-catch linker |
| SELEX | systematic evolution of ligands by exponential enrichment |
| SpAM | spatially arrayed mixtures |
| SPIMS | solid phase *in situ* mass spectrometry |
| SPOS | solid phase organic synthesis |
| SPS | solid phase synthesis |
| SURF | synthetic unrandomization of randomized fragments |

| TBTU | 2-(benzotriazol-1-yl)-1,1,3,3-tetramethyluronium tetrafluoroborate |
|---|---|
| TFA | trifluoroacetic acid |
| TFP | tri-2-furylphosphine |
| TGF-$\beta$ | transforming growth factor $\beta$ |
| THF | tetrahydrofuran |
| THP | tetrahydropyran |
| TLC | thin-layer chromatography |
| TMAD | tetramethylazodicarboxamide |
| TOCSY | total correlation spectroscopy |
| TOF–SIMS | time-of-flight secondary ion mass spectrometry |
| TPAP | tetra-*n*-propylammonium perruthenate |
| UV | ultraviolet (spectroscopy) |
| VL | virtual library |
| VLSI | very large scale integrated |
| VLSIPS | very large scale immobilized peptide synthesis |

# 1 Combinatorial synthesis

## 1. Introduction

Combinatorial chemistry has its earliest origins in solid phase peptide synthesis. Merrifield's Nobel Prize-winning invention depended on the use of consistent and reliable reaction conditions for peptide couplings, and the use of a polymeric solid phase to permit the simple separation of products from reagents (Merrifield 1963). The solid phase techniques have been greatly refined over the years to allow the creation of essentially any peptide molecule in a consistently high yield. In the mid 1980s, several academic teams realized that so reliable and consistent were the synthetic reactions, that the same conditions could be used to make many peptides simultaneously in the same reaction vessel. In this way, productivity could be enhanced, and the discovery of peptide epitopes responsible for protein recognition and binding could be accelerated. In particular, Houghten's use of 'tea-bags' as porous containers for solid phase resin beads, allowed the same peptide coupling step to be applied to many beads simultaneously irrespective of the sequence already attached to the bead (Houghten 1985). However, the mix and split procedure, first described by Furka (Furka *et al.* 1988), and used shortly after by Houghten (Houghten *et al.* 1991) and Lam (Lam *et al.* 1991), allowed huge numbers of peptides to be made in a very few number of chemical steps, and combinatorial chemistry was born.

In essence then, combinatorial chemistry is based on efficient, parallel synthesis, in that many more chemical compounds can be generated in a library than the number of steps used in the synthesis. This is a complete contrast to synthetic organic chemistry as it has been practised for the last 100 years, as traditional chemistry is usually distinguished by use of several synthetic steps in the preparation of one compound.

The impact of 'combichem' on both academics as well as industrial chemists has been conceptual as much as practical. The early papers in this field challenged the long-held beliefs that all synthetic chemical compounds should be made individually, in a fully purified and characterized state. Many groups have since explored high speed chemistry, both of mixtures and of single compounds, for biological applications such as drug discovery, or receptor–ligand studies, although supramolecular chemistry has also benefited. The scope of chemistry employed has branched well beyond peptide synthesis to include other key chemical transformations and structural classes. Some groups have preferred to use solution phase library chemistry (see Chapter 4) for library synthesis as this generally requires much less development and validation, but still the majority of compound libraries are made on a solid phase of some description (see Chapters 2 and 3). Resin beads are still the preferred vehicle, but grafted polymeric 'pins', paper or polymer sheets, and even glass chips have all been used.

From the outset, one major distinction can be made between mixtures and single compounds. Compound mixtures can present a highly effective method for the discovery of compounds with particular biological or supramolecular activity, as long as there is a suitable reporting method to permit the identification of the preferred compound or compounds in the mixture. It is not possible however, to find compounds with any particular bulk property, such as magnetic or semiconducting properties, from testing mixtures. If compounds with specific bulk properties are required, it is a prerequisite to make and test isolated compounds in a pure form. Thus combinatorial chemistry methods to make discrete compounds have generally found the widest applications, whereas mixtures have mainly been applied in the search for biological activity.

With the increased productivity afforded by these new combinatorial techniques, the challenge has transferred to the efficient screening of all these products. Different groups have generated various solutions to this problem, with many elegant library designs to permit unambiguous identification of the most biologically active products. Whilst the iterative deconvolution of mix and split libraries is widely used, much invention has been applied to the use of library encoding methods. In applying this methodology, a simple chemical or analytical analysis of a tag molecule can unequivocally define the active test compound structure (Chapter 5). Encoding methods can be chemical but also they can be electronic, with the use of transponder devices that can be written and read to find the identity of preferred library components.

Thus, it can be seen that combinatorial chemistry is not one technique. It embraces a diversity of chemistry techniques: in solution or on solid phase, making libraries in mixtures or as single compounds. It may rely on esoteric encoding methodologies, or it may use none. The compounds may have been made manually, or by using the most sophisticated laboratory automation.

In addition to revolutionizing the way chemistry is now viewed and having a major impact on methods for the discovery of biologically active compounds, combinatorial chemistry has also challenged some of the science's most inventive minds to generate elegant and practical solutions to real technological problems. This book hopes to convey a sense of the diversity and invention in this area, and to illustrate how any chemist might apply some of the methods to their own work.

## 2.   The drug discovery process

The outcome of the drug discovery process is the identification of a chemical structure that has both the desired potency against a nominated biological target, and also has suitable bioavailability and efficacy in an appropriate animal model of the targeted disease. It is this structure that the discoverer ultimately protects through a patent filing. Drug discovery is both a lengthy and expensive business for it frequently takes five years from initiating a project to the point where a potential drug is nominated for development and clinical trials.

One reason for this lengthy discovery period is that within this process, the synthesis of exploratory compounds can often be the slowest step. Early on in the lifetime of a drug discovery project medical chemists need to find a lead compound—a structure with some degree of affinity, however small, for the biological target. With this lead in hand they then proceed to the second phase: the identification of a drug development candidate by the stepwise, incremental improvement of the lead's structure (Fig. 1.1).

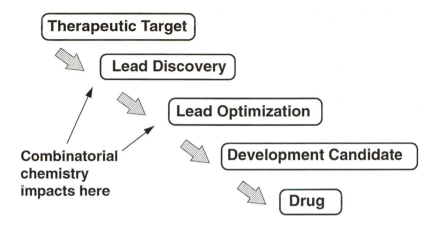

**Fig. 1.1**  The key steps in the drug discovery process and the steps that can be influenced by the application of combinatorial chemistry.

To address the first phase, the drug discoverer often uses a known literature or competitor's compound as a lead. But often, the target has no known ligand, and under these circumstances empirical screening may be the only approach for lead discovery. Historically, the main source of biologically active compounds for use in drug discovery programmes has been natural products, isolated from plant, animal, or fermentation sources, and the past success stories of the discovery of $\beta$-lactams, tetracyclines, avermectins, and taxol are well known. However, despite the broad range of natural product structural diversity, finding activity for a specific biological target is nearly always a highly challenging process, and companies usually resort instead to screening the compound files that they have built up over many years of drug discovery.

The average pharmaceutical company archive contains in the region of 200 000 compounds. To allow the assessment of such large numbers in a reasonable time-scale, automated methods for rapid screening have been required to allow testing of thousands of compounds per week. However, the flaw in this approach is that the structural diversity of company compound collections is limited, biased by the range of structures previously synthesized by that company. It is a moot point whether the frequent discovery of leads based on specific structural classes, benzodiazepines for example, is a consequence of their structural attributes, or whether their occurrence reflects the disproportionate number of these structures held in company vaults. The same argument could be applied to dihydropyridines (valued for their calcium channel blocking properties) or $\beta$-lactams.

We have reached a stage in the development of drug discovery where novel receptors and enzymes are increasingly being identified as therapeutic targets, but drug discoverers are restricted to screening natural product extracts or moribund compound files in order to find that elusive hint of activity that could launch a new synthetic programme. Thus any technique that could lead a new path out of this structural desert, and permit the rapid synthesis of thousands or even millions of new compounds for lead discovery would quickly capture the imagination of pharmaceutical chemists. Combinatorial chemistry provides such an opportunity, as it is a technique that in its power to generate huge numbers of new compounds is already changing the face of drug discovery.

It has been estimated that the total number of possible small organic molecules is in the region of $10^{200}$ different structures (Czarnik 1995). With this total collection it might be assumed that a drug for any target may be readily discovered. But when one considers that accumulating just one molecule of each of these compounds would result in a quantity of material that exceeds the mass of the universe by a factor of $10^{128}$, it can be appreciated that even empirical screening is limited by the number of compounds that realistically can be synthesized and tested.

Lead discovery from combinatorial libraries is primarily a speculative process: it has, perhaps unfairly, been called 'irrational' drug design. Initially there may be no attempt to design an active library of compounds, although library design is becoming a key factor in the synthesis of quality libraries. Rather, it is the number and variety of structures that libraries can offer that are their main attraction. Often, any structural preconceptions of what might, or should have, affinity for the target protein are usually ignored in favour of the serendipitous discovery of a novel lead. This isn't to suggest that rational design has no place in combinatorial chemistry, as any knowledge of the receptor or enzyme structure, however limited, can be used to design a library of compounds for a specific target.

Lead discovery is only one place where combinatorial chemistry is impacting on drug discovery. The second phase of taking the lead and improving its profile to the point where a drug development candidate can be nominated is also positively affected by the new high speed synthesis techniques. The chemist proceeds by formulating and testing compound binding hypotheses by synthesizing novel compounds for screening. Our understanding of the mechanisms by which compounds bind to a biological target is poorly developed, and often it is difficult to modify the lead compound in a way that reflects a rational understanding of the compound's mode of binding to the protein target. Chemists are finding that combinatorial chemistry offers a way by which many compounds can be synthesized in parallel to optimize a lead compound's activity in the absence of any binding model.

Even if the medicinal chemist has an understanding of how the lead compound works, combinatorial chemistry provides a way of rapidly exploring structure–activity relationships (SAR). For example, the chemist may recognize the need for a lipophilic side-chain on his lead molecule. Whereas in the past he may have had the time only to synthesize compounds containing methyl, propyl, and benzyl substituents, combinatorial chemistry can be used to make 100 or more analogues to map out the potential range of this substituent, allowing a more thorough investigation and possibly identifying the unexpected active analogue.

Combinatorial chemistry has changed the way we approach synthetic chemistry. In industrial environments especially where there is pressure to discover new, commercially profitable agents as quickly as possible, these techniques have already made a considerable impact. Nearly every pharmaceutical company has now established at least one group working in this area, and we have seen the emergence of huge numbers of small, specialized biotechnology companies offering new methods and applications of combinatorial synthesis.

The commercial applications of combinatorial chemistry have spawned new branches of the pharmaceutical and biotechnology industry. However, the area is not only useful commercially; it also offers an enormous intellectual challenge for industrialists and academics alike. The remainder of this chapter sets the scene for understanding this challenge by defining the basic concepts of combinatorial chemistry.

## 3.  What is combinatorial synthesis?

The essence of combinatorial synthesis is the ability to generate large numbers of chemical compounds very quickly. This one key advantage has inspired many different uses, employing many variants of the technology. Combinatorial chemistry has also initiated a reassessment of the traditional methods of organic synthesis. Chemistry in the past has been characterized by slow, steady, and painstaking work; combinatorial chemistry has broken many of the preconceptions and permitted a level of chemical productivity undreamed of just ten years ago.

The speed of combinatorial synthesis has come about by discarding many of the dearly held precepts of organic synthesis; that all compounds and intermediates need to be fully purified and characterized. Instead by using methods that use reliable chemistry and simple but effective purifications, combinatorial chemistry has allowed great productivity although sometimes at the expense of quality. As we shall see, the pendulum is now swinging back to an equilibrium where quality becomes an equal partner to quantity, as it is no longer necessary to tolerate poor quality compounds even when made in large numbers.

How have these steps forward in productivity been taken? In the past chemists have made one compound at a time, in one reaction at a time. For example compound A would have been reacted with compound B to give product AB, which would have been isolated after reaction work-up and purification through crystallization, distillation, or chromatography (Fig. 1.2). In contrast to this approach, combinatorial chemistry offers the potential to make every combination of compound $A_1$ to $A_n$ with compound $B_1$ to $B_n$. The range of combinatorial techniques is highly diverse, and these products could be made individually in a parallel fashion or in mixtures, using either solution or solid phase techniques. Whatever the technique used the common denominator is that productivity has been amplified beyond the levels that have been routine for the last hundred years.

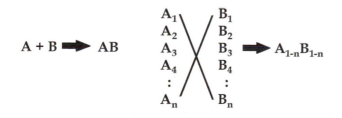

**Fig. 1.2**   The contrast between orthodox chemistry where products are prepared individually and combinatorial chemistry that offers the potential to prepare all the products of a range of starting materials.

In particular, the types of chemistry employed have been biased towards methods that can be expected to give good yields and purities, and furthermore there has been a reliance on techniques such as solid phase chemistry that do not require reaction work-up and also offer a degree of purification through simple physical separations such as filtration.

The next chapter describes how combinatorial chemistry evolved out of the earliest solid phase synthesis of peptides, and how simple techniques have been developed to achieve a remarkable growth in synthetic productivity.

## References

Czarnik, A.W. (1995). Why combinatorial chemistry, why now (and why you should care). *Chemtracts: Org. Chem.*, **8**, 13–18.

Furka, A., Sebestyen, F., Asgedom, M., and Dibo, G. (1988). *Abstr. 14th Int. Congr. Biochem. Prague, Czechoslovakia*, **5**, 47.

Houghten, R.A. (1985). General method for the rapid solid-phase synthesis of large numbers of peptides: Specificity of antigen-antibody interaction at the level of individual amino acids. *Proc. Natl. Acad. Sci. USA*, **82**, 5131–5.

Houghten, R.A., Pinilla, C., Blondelle, S.E., Appel, J.R., Dooley, C.T., and Cuervo, J.H. (1991). Generation and use of synthetic peptide combinatorial libraries for basic research and drug discovery. *Nature (London)*, **354**, 84–6.

Lam, K.S., Salmon, S.E., Hersch, E.M., Hruby, V.J., Kazmierski, W.M., and Knapp, R.J. (1991). A new type of synthetic peptide library for identifying ligand-binding activity. *Nature (London)*, **354**, 82–4.

Merrifield, R.B. (1963). Solid phase peptide synthesis. I. The synthesis of a tetrapeptide. *J. Am. Chem. Soc.*, **85**, 2149–54.

# 2  Library synthesis on resin beads

## 1.  Solid phase chemistry

The synthesis of combinatorial libraries usually employs solid phase chemistry. The recent renaissance and growth of interest in solid supported chemistry stems from the opportunity to use filtration as a separation technique to purify both intermediates and final compounds. The ability to synthesize compounds on a resin bead for example, to force the reaction to completion by the addition of excess reagents and monomers, and then being able to remove all the unwanted materials by a simple filtration and wash is at the heart of many library syntheses (Fig. 2.1).

This is not to say however, that solution methods are no longer suitable for combinatorial libraries. With appropriate chemistry, either that which is high yielding or that generates products which allow simple liquid–liquid extraction, solution phase library synthesis is highly feasible (see Chapter 4) and indeed has been chosen as a preferred strategy by many academic and industrial groups.

The techniques for solid phase synthesis (SPS) are based extensively on the pioneering work of Merrifield (1963) who, in a paper that unusually for such a significant publication bore his name as sole author, described the use of substituted resins as the solid phase for synthesis of peptides. Although the expression 'solid phase synthesis' was coined at this time, it is not an accurate description. The synthesis of peptides does not occur on the surface of a truly solid material, but instead within a gel-like matrix of connected polymeric molecules swollen by the ingress of solvent molecules. In fact rather than on a solid phase, the peptide is synthesized in a highly fluid, almost solution-like, environment.

Solid phase synthesis naturally lends itself to the production of peptides, because of the limited range of synthetic transformation that are required for

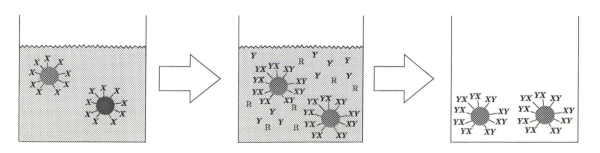

**Fig. 2.1**  The synthesis of the dimer XY on resin beads is facilitated by using excess monomer Y and reagents, R, which can be removed by a simple filtration at the end of the synthetic step.

their synthesis. Each of the key reactions in the process, indeed every nuance of every step in peptide synthesis, has been scrutinized and optimized to allow the production of peptides of sizeable length in high overall yields. A consideration of the mathematics involved demonstrates that for high overall yields, individual steps need to proceed in essentially quantitative yields. If each coupling cycle proceeds in a 99% yield, a ten amino acid peptide will be produced in around 90% overall yield. Longer sequences or lower chemical yields will be reflected in lower recovery and purity.

Early on in the development of solid phase methods some scientists subjected this approach to major criticisms. First they proposed that the likelihood of this being a viable approach to peptide synthesis was very low indeed, and that such consistently high chemical yields were unattainable. Fortunately, they were wrong, and the synthesis was improved, ultimately to the point where automated peptide synthesis was made possible. There are now a range of commercially available peptide synthesizers, all reliant on the latest perfected incarnation of the original Merrifield concept of peptide synthesis. Secondly, many chemists considered that the concept was a 'sloppy' approach to synthesis as intermediates were not isolated, purified, or characterized. Intriguingly, these are criticisms that 30 years later have also been directed at combinatorial chemistry—evidence that in chemistry some attitudes never change!

Given the reliability and generality of solid phase peptide synthesis, it was an obvious extension that the earliest combinatorial chemistry was the synthesis of peptide libraries. We shall see however that the range of solid phase chemistry has expanded enormously beyond peptides in recent years to permit the combinatorial synthesis of an increasing number of desirable and diverse drug-like structures. The strictures of the critics still apply. It is essential for all types of solid phase library chemistry that the yields are high to avoid an accumulation of unwanted by-products. If this cannot be achieved, and very little chemistry even remotely approaches peptide synthesis in efficiency and generality, then methods for separation of products from contaminants must be worked into the library synthesis.

## 2.   Resin beads

The use of solid supports for organic synthesis relies on three interconnected requirements:

(a)   A cross-linked, insoluble, but solvent swellable polymeric material that is inert to the conditions of synthesis.

(b)   Some means of linking the substrate to this solid phase that permits selective cleavage of some or all of the product from the solid support during synthesis for analysis of the extent of reaction(s), and ultimately to give the final product of interest.

(c)   A successful synthetic procedure compatible with the linker and the solid phase.

In practise, the most useful applications of solid phase synthesis have used small (80–200 $\mu$m) spherical resin beads. These can be prepared in the laboratory, but are generally sourced from commercial suppliers who make them on an industrial scale. Preparation typically involves the addition and dispersion of an organic phase of monomer and cross-linker in an aqueous solution. A free radical initiator is dissolved in the organic mixture, so that raising the temperature starts the polymerization to convert the microdroplets to solid spherical resin beads. As the polymers grow their insolubility rises, so to prevent premature precipitation a solution stabilizer is usually added to the aqueous mixture. The process is thus carefully monitored to control the rate of polymerization and the size of particle. After completion of polymerization, the beads are collected by filtration and washed free of unreacted monomers and the aqueous phase. After thorough drying, the consistency of size is ensured by sieving.

## 2.1   Cross-linked polystyrene

The resin bead first used by Merrifield, and still widely used today, is a gel-type polymer made from cross-linked polystyrene. During the polymerization process, around 1% divinylbenzene is added to the styrene to link the forming polystyrene chains together. This degree of cross-linking is sufficient, by holding the chains together, to give mechanical strength and insolubility—the resin bead is essentially one gigantic macromolecule—but not so much cross-linking as to prevent the swelling of the bead when immersed in solvent or as the organic compounds grow on the polymer (see Fig. 2.2). Indeed swelling is an essential feature of gel resins, as it reflects an internal flexibility of polymer backbone that can move to maximize the available functionality as well as permitting free diffusion of solvents and reagents into the bead. Although used for a wide range of chemistry, the polystyrene clearly has limitations if used for highly electrophilic reagents, and the thermal stability of the resin limits reaction conditions to below 130 °C.

An observation made by workers at Parke-Davis and the University of Edinburgh is that polystyrene beads are not always perfectly inert during solid phase synthesis (MacDonald *et al.* 1995). Analysis of the final library products by both [1]H NMR and MS revealed the presence of impurities arising from

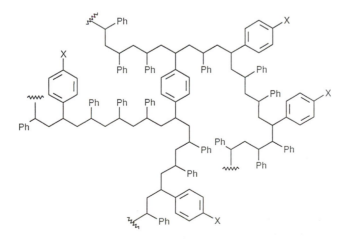

**Fig. 2.2**   The internal molecular structure of polystyrene. The groups marked X could be any suitable functionality, but are generally derived from the chloromethyl group.

both the compound synthesis and the solid phase itself. Impurities in commercial resin itself stemming from trapped solvents or products of incomplete polymerization are a significant problem as they can reduce the functionality available for loading and will contaminate the library products.

Various methods for avoiding or removing the contamination were examined. Successive pre-washing of the resin beads with a broad range of solvents followed by drying had an appreciably beneficial effect on the quality of final compounds, although it was noted that the quality of resins varied greatly between the various commercial suppliers. Additionally, exposure of the resins to the cleavage conditions generated a further number of impurities. Pre-treatment of the resins with these reagents did remove a proportion of such impurities, but they appeared again following repeated treatment due to a gradual degradation of the polystyrene. It was concluded that although pre-washing was advantageous, purification of the final library products was the only method that guaranteed removal of impurities.

During the process of solid phase synthesis, molecules are attached and further derivatized whilst attached to the polymer backbone. This is achieved by the initial presence of chemical functionality within the polymer. Such functionality can either be part of the monomer prior to polymerization, or can be introduced through chemical modification of the polymer beads. The latter route offers a more flexible strategy as it permits the introduction of a range of different linkers to the polymer, and also minimizes any loss of functionality buried in inaccessible places within the bead.

Merrifield resin

**Fig. 2.3**   Loading of Merrifield chloromethyl resin with *N*-protected amino acids to generate an ester linkage to the solid phase.

The original Merrifield resin was prepared by chloromethylation of the benzene rings, and attachment of *N*-protected amino acids was achieved by the nucleophilic displacement of the chloride by a carboxylate anion (Fig. 2.3). The resulting ester bond was stable to the conditions of peptide synthesis which proceeded by the stepwise addition of *N-tert*-butyloxycarbonyl (Boc)-protected amino acid monomers using dicyclohexylcarbodiimide as the coupling reagent. In improving the process of solid phase peptide synthesis, many other superior coupling reagents have been developed since Merrifield's original work. Side-chain functionality on the amino acids was usually protected with benzyloxycarbonyl or benzyl groups that remain intact until the end of the synthesis. Synthesis of the peptides proceeded from the C-terminal to the N-terminal, as this avoided the risk of amino acid racemization. Deprotection of the Boc groups using trifluoroacetic acid (TFA) preceded addition of the next amino acid residue, and the peptide was finally cleaved from the polystyrene and the side-chains deprotected under vigorous acidic conditions using anhydrous liquid HF. Merrifield's ingenuity was to use the resin bead both as a protecting group for the carboxy terminus as well as providing a solid vehicle to aid purification.

Many alternative linking groups on polystyrene have been described, generally at a functionality level of 0.5–1.2 mmolg$^{-1}$ of resin, and these linkers permit the release of carboxylic acids under more mild conditions (see Chapter 7, Section 1.1).

One concern often expressed in solid phase synthesis is the possibility of site–site interactions between molecules in the bead. Scott *et al.* (1977) indicated that even if polystyrene rigidity was enhanced by increasing cross-linking to 2%, significant site–site interactions could still take place. It was thus suggested that peptide synthesis should take place at very low loadings to minimize interactions between the growing peptide chains. Although gel polymers swell to accommodate the growing peptide, the hydrophilic nature and hydrogen bonding potential of peptides, contrasted with the hydrophobic nature of the polystyrene matrix, almost certainly guarantees peptide–peptide chain interactions at the expense of effective synthesis. The hydrophobicity of the polystyrene induces chain folding in which the peptide satisfies its own hydrogen bond requirements rather than being solvated and this severely limits synthetic access to the exposed end of a growing peptide chain. Furthermore, solvents that expand polystyrene would have a tendency to collapse peptides, whilst solvents such as DMF that expand the peptide have non-optimal polystyrene swelling properties. Ideally, both polymer and peptide should be equally expanded under reaction conditions to enhance the rate and yields of synthesis. These issues led to the development of new polymer supports for peptide synthesis.

## 2.2   Polyamide resins

As a consequence of the concerns listed above, Sheppard designed polyacrylamide polymers for peptide synthesis as it was expected that these polymers would more closely mimic the properties of the peptide chains themselves and have greatly improved solvation properties in polar, aprotic solvents (e.g. DMF, or *N*-methyl pyrrolidinone) (Atherton *et al.* 1975; Arshady *et al.* 1981; Atherton and Sheppard 1989). A favoured combination was *N,N*-dimethyl acrylamide for the backbone, cross-linked with *N,N'-bis*-acryloylethylenediamine and functionalized through *N*-acryloyl-*N'*-Boc-*β*-alaninylhexamethylenediamine (Fig. 2.4). This resin swells in polar solvents but has limited ability to swell in less polar solvents such as methylene chloride.

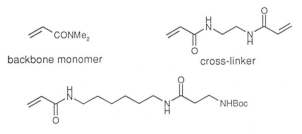

backbone monomer

cross-linker

backbone monomer with protected functional group

**Fig. 2.4**   The precursors used in the preparation of polyacrylamide resins. The functionalized backbone monomer provides a protected amine handle for building the peptide.

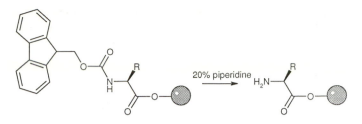

**Fig. 2.5**  Protection of amino acid monomers with the Fmoc group allows ready cleavage with 20% piperidine.

In combination with the design of a novel, polar solid phase support, Sheppard also proposed the use of a new protection and linking strategy for peptide synthesis (Atherton *et al.* 1981). Instead of the Merrifield peptide synthesis approach that depended on a benzyl ester linkage and Boc protection for the amino acids, more mild protection/deprotection/cleavage methods were sought. It had been known that the repetitive cleavage of the Boc protecting groups using TFA, which often took 30 minutes to complete, and the final vigorous benzyl ester linker and side-chain protection cleavage, usually with liquid HF, led to significant decomposition of the peptides. Sheppard examined a range of other amine protecting groups for the amino acid monomers, focusing especially on base-labile groups. In this way, the mild alkaline cleavage conditions would not affect the integrity of the peptide, and strong acid could be reserved for the final linker cleavage and side-chain deprotection. The protecting group finally chosen was the fluorenylmethoxycarbonyl (Fmoc) group, which although stable to the methods of peptide synthesis, could be readily removed within minutes by treatment with 20% piperidine (Fig. 2.5). The recommended acid-labile side-chain protection was Boc for amines and *tert*-butyl ethers and esters. The solid phase linker employed was the hydroxymethylphenoxyacetic acid (HMPA) attached through a diamine to the polyamide resin backbone as this presents a highly acid-labile ester linkage.

The combination of the Fmoc protecting group and the Boc side-chain protection has become the standard method for the solid phase synthesis of peptides, and consequently is also widely used in the preparation of peptide and some non-peptide combinatorial libraries. However, this protection strategy is not used solely with the polyacrylamide resin, and is to be commonly found used with cross-linked polystyrene resin. The equivalent of the HMPA linker used with polystyrene is commonly known as Wang resin (see Chapter 7, Section 1.1).

Other polyamide-based resins have been designed using alternative backbone monomers. In particular, replacing the *N,N*-dimethyl acrylamide with the more lipophilic *N*-acryloyl pyrrolidine produces a polymer that swells in solvents such as methanol, ethanol, 2,2,2-trifluoroethanol, isopropanol, acetic acid, and water; solvents which generally do not swell polystyrene sufficiently for synthesis (Stahl *et al.* 1980), but in addition it also swells well in methylene chloride.

## 2.3   Controlled pore glass

Polystyrene and polyacrylamide are resins designed for batchwise solid phase synthesis. They generally lack the structural rigidity required for continuous flow synthesis purposes where the resin needs to be permanently solvated. Consequently there have been a large number of new macroporous polymers designed with rigid open pores that permit a ready and continuous solvent flow. However, as combinatorial chemistry by its nature is a batchwise procedure, these resins are not discussed further here. The only exception to this is the use of controlled pore glass (CPG), a rigid, glass-derived bead material, compatible with any type of solvent, stable to aggressive reagents and extremes of pressure and temperature, that has occasionally been used for the combinatorial synthesis of peptides and oligonucleotides (Köster *et al.* 1983).

## 2.4 TentaGel resin

A further type of polymer support has had a significant impact on combinatorial solid phase synthesis. As with the polyacrylamide resins, in order to produce a polar reaction milieu that is closer to the solvents generally used by solution synthetic chemists, grafted polymer beads have been prepared. The most pre-eminent of these is TentaGel resin (Rapp Polymere Gmbh) (Grubler *et al.* 1994) which consists of polyethylene glycol attached to cross-linked polystyrene through an ether link, and combines the benefits of the soluble polyethylene glycol support with the insolubility and handling characteristics of the polystyrene bead. The resin was originally prepared by the polymerization of ethylene oxide on cross-linked polystyrene already derivatized with tetraethylene glycol to give polyethylene glycol chains (Fig. 2.6) (Bayer *et al.* 1985). Optimized TentaGel grafted resins generally carry polyethylene glycol chains of about 3 kDa in size, accounting for about 70–80% of the beads by weight. An alternative form, marketed as PEG-PS (Perseptive), has the polyethylene glycol chains linked through an amide bond.

It is remarkable that the cross-linked polystyrene backbone is sufficiently flexible to accommodate the polyethylene glycol and flex further still to permit the synthesis of peptides or other organic molecules. However, it should be noted that in general, the loading density on TentaGel (0.25 mmolg$^{-1}$) is lower than that obtainable on standard cross-linked polystyrene (usually in the range 0.5–1.2 mmolg$^{-1}$).

The synthetic environment within TentaGel is closely related to ether and tetrahydrofuran solvents, and consequently has the potential for compatibility with the large range of reactions that are currently being investigated for compound library synthesis (Rapp *et al.* 1990, Rapp and Bayer, 1994). For example, work on the synthesis of $\beta$-turn mimics on solid phase could only be completed on TentaGel, as polystyrene beads did not permit solvation by

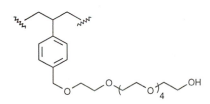

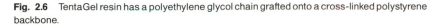

**Fig. 2.6** TentaGel resin has a polyethylene glycol chain grafted onto a cross-linked polystyrene backbone.

aqueous solvent comixtures (Virgilio and Ellman 1994). The hydrophilic nature of the resin additionally facilitates release of products, or screening on the bead, in an aqueous environment.

Studies with a confocal microscope to examine the penetration of fluorescein-labelled porcine pancreatic trypsin, an enzyme complex with a molecular weight of 23.8 kDa, revealed that 90 $\mu$m TentaGel resin beads allow enzymes readily to access the centre of the polymer (Quarrell *et al.* 1996). Whilst this seems an improbable event, it should be recalled that the polymer is exceptionally flexible, and furthermore is approximately 10 000 times the diameter of the enzyme. TentaGel resin provides an internal environment that does not interfere with the normal behaviour of enzymes in aqueous solution, and the assay of compounds whilst attached to the bead will be less affected by the presence of the polymeric support material.

## 2.5   Magnetic beads

The various beads described above are usually handled by manual methods, but in the search for improved procedures for distributing resins, new procedures have been proposed. One method frequently employed is the production of an isopycnic suspension, where the pipetting of slurries is improved by ensuring that the resin and solvent have similar densities. One other approach to the manipulation of resin beads is the development of the magnetic polymer bead (Szymonifka and Chapman 1995).

Polydivinylbenzene was nitrated and the nitro groups reduced with ferrous sulfate hexahydrate. This reduction produces ferrous and ferric ions within the bead that could be converted into magnetite crystals by the addition of concentrated ammonium hydroxide solution followed by gentle heating. After thorough washing, the beads contained 24–32% iron by weight and were easily manipulable by a hand-held bar magnet. These beads were used for the synthesis of a protected dipeptide (Fmoc–Phe–Ala–OH) but the use of these beads has yet to be taken up by other scientists in the combinatorial chemistry area. It is likely that they are seen as unattractive as the heavily cross-linked polymer will not permit high levels of chemical functionality, and as a warning footnote points out, the iron in the bead is reactive under a number of synthetic conditions.

## 3.   Speeding up peptide synthesis

In the mid 1980s, although the techniques for the solid phase synthesis of single peptides were well developed, there was a limit to the numbers of compounds that could conveniently be made. However, there were an increasing number of applications, primarily the studies of protein–protein recognition events, that required large numbers of different peptide sequences. To overcome this synthetic bottleneck a number of new techniques were developed that allowed hundreds of peptides to be made in just a few weeks. One such technique was the Multipin apparatus (see Chapter 3, Section 1), and another was the mesh packet or 'tea-bag' designed by Houghten (1985).

The polypropylene mesh bag, with dimensions of approximately 15 × 20 mm, was filled with resin beads, sealed, and labelled with a black marking pen to allow easy later identification. The 'tea-bag' mesh size of 74 $\mu$m was too small to allow resin beads to escape, but solvents and soluble reagents could readily enter. The principle of their use was to make multimilligram (up to 500 $\mu$moles) quantities of a single peptide sequence in each packet—sufficient for full characterization and screening. But to save time and work when making many peptides simultaneously, bags could be combined into the same reactors for common chemical steps. For example, in the synthesis of 40 different peptides, all the bags were initially charged with resin beads bearing a Boc-protected amino acid, and then either with the use of peptide synthesizers or by a completely manual procedure, the packets were combined for resin deprotection, washing, and neutralization steps. The bags were then sorted into groups for the addition of the next amino acid. After this addition step, the bags could be recombined for washing, deprotection, and neutralization. After the appropriate number of coupling steps the individual bags, each bearing a unique peptide sequence were subjected to the conventional HF/anisole procedure to cleave the peptide from the beads using an apparatus designed to treat 20 bags at once (Houghten *et al*. 1986).

The methodology was used for the synthesis of 247 replacement analogues and 13 identical controls of a 13 amino acid sequence (residues 98–110) of the influenza haemagglutinin protein (HA1). These 260 peptides were made in four weeks, and obtained in 10–20 mg amounts. Importantly, analysis of the peptides by HPLC and amino acid analysis revealed them to be as good in purity as peptides prepared using free resin synthesis. Studies of the binding of these peptides to monoclonal antibodies to the sequence 75–110 of HA1 demonstrated that the aspartic acid at position 101 was of critical importance to the binding interaction.

By this method of combining common reaction steps, considerable time could be saved in preparing many different peptides. With hindsight this is a classic example of a combinatorial synthesis, as the intention was clearly to improve synthetic efficiency. The further innovation that allowed synthetic productivity of peptides to really accelerate, by the use of resin mix and split procedures, is described in the next section.

## 4. Mix and split library synthesis

The impact of combinatorial chemistry stems from the opportunity it provides to rapidly and efficiently generate huge numbers of peptides or other compounds. The technique that had the greatest initial impact on this productivity is a method described as *mix and split* synthesis (other names are applied to the process including 'split synthesis', 'one bead–one compound', 'split-and-mix', the 'Selectide' process, 'portion mixing', and 'divide-couple-recombine', but these all refer to essentially the same methodology). This technique was pioneered by Furka (Furka *et al*. 1988, 1991) and has been enthusiastically adapted by many others since its first disclosure.

As an example, to make 10 000 compounds using orthodox chemistry would normally require 10 000 reaction vessels. If the products were trimers prepared through three synthetic steps, there would be at least 30 000 separate chemical reactions required. This is a necessary prerequisite for making that number of compounds. The mix and split approach is much more efficient. By combining different substrates requiring the same reaction conditions into the same reaction vessel, making the same 10 000 trimeric compounds can be done in as little as the cube root of 10 000 or approximately 22 reaction vessels.

**Table 2.1**   The potential size of combinatorial libraries generated for varying lengths of product and numbers of monomers

| Number of monomers used | Total number of compounds in the library | | | | |
| --- | --- | --- | --- | --- | --- |
| | Dimers | Trimers | Tetramers | Pentamers | Hexamers |
| 3 | 9 | 27 | 81 | 243 | 729 |
| 5 | 25 | 125 | 625 | 3125 | 15 625 |
| 10 | 100 | 1000 | $10^4$ | $10^5$ | $10^6$ |
| 20 | 400 | $8 \times 10^3$ | $1.6 \times 10^5$ | $3.2 \times 10^6$ | $6.4 \times 10^7$ |
| 100 | $10^4$ | $10^6$ | $10^8$ | $10^{10}$ | $10^{12}$ |

The mix and split process relies on the synthesis of the library compounds on small resin beads. A quantity of this support material is divided into a number of equal portions and each of these portions are individually reacted with a different monomeric starting material. For peptide libraries the monomer will be an amino acid suitably protected, usually at the N-terminus, with a Boc or Fmoc group. The coupling reaction is usually pushed to completion by the addition of excess monomer and coupling reagent; this excess is not a problem as it can be removed from the solid phase by subsequent washing. The individual portions of resin are recombined, the solid phase is thoroughly mixed, and may then again be divided into portions. Reaction with a further set of reagents (usually following a deprotection step to reveal the next key functional group) gives the complete set of possible dimeric units as mixtures. This whole process may then be repeated as necessary (for a total of $n$ times). The number of compounds obtained arises from the geometric increase in products: in this case $x$ (monomers) to the power of $n$ (total number of coupling steps in the synthetic procedure used to make the compounds).

Table 2.1 describes how the numbers of compounds prepared through the mix and split procedure is dependent on the length of the oligomers being produced and the number of monomers used for each position of the oligomers. The library sizes quoted in the table are based on the assumption that each monomer position in the products contains the same set of precursors, but this is by no means a necessity.

Whilst peptide chemists will often use all 20 naturally occurring amino acids in each position (although in reality 19 only are usually employed, as cysteine often generates problems through oxidation of the sulfide), as combinatorial chemists have investigated non-peptidic structures the numbers and types of monomer in each position have exhibited a greater diversity. Furthermore, the products made do not necessarily need to be linear. Suitable templates with two, three, or more functional groups allow the production of combinatorial libraries of branched products using any type of currently available solid phase chemistry (Fig. 2.7).

Some workers (Pirrung 1993) have specifically described this type of library as 'combinatorial' preferring to use the description 'permutational' for oligomeric (e.g. peptide) libraries.

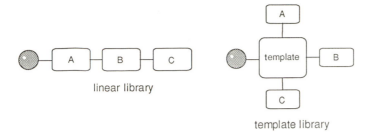

**Fig. 2.7** The constituent compounds in combinatorial libraries can either be linear or derived from the derivatization of a multifunctional template. In each case the mathematics of library size is the same.

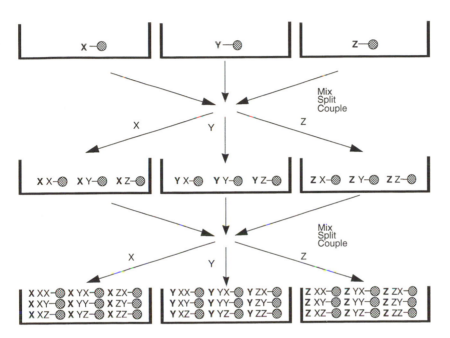

**Fig. 2.8** The mix and split synthetic scheme for the synthesis of a 27 member trimer library. The final compounds are generated in three mixtures each containing nine compounds.

The mix and split process is illustrated schematically in Fig. 2.8 for a simple example of a $3 \times 3 \times 3$ library, giving all 27 possible combinations of trimeric products. X, Y, and Z could be nucleotides, in which case the product would be an trinucleotides, or amino acids in which case the final products would be tripeptides.

It can be seen that this mix and split procedure finally gives three mixtures consisting of nine compounds each, and there are now several ways of progressing these compounds to biological screening. Although they can be tested whilst still attached to the bead, a favoured method is to test the compounds as a mixture in solution following cleavage from the solid phase. Activity in any given mixture reveals the partial structure of active compounds within the library, as the residue coupled last (usually the N-terminal residue in a peptide library) is unique to each mixture. Identification of the most active

**Table 2.2**  The iterative determination of the most potent peptide sequences to bind to the 19B10 monoclonal antibody

|  | Peptide sequence | IC$_{50}$ ($\mu$M) |
|---|---|---|
| 1st round | Ac–DVXXXX–NH$_2$ | 250 |
|  | Ac–DIXXXX–NH$_2$ | 318 |
|  | Ac–DMXXXX–NH$_2$ | 752 |
| 2nd round | Ac–DVPXXX–NH$_2$ | 41 |
|  | Ac–DVEXXX–NH$_2$ | 146 |
|  | Ac–DVQXXX–NH$_2$ | 215 |
| 3rd round | Ac–DVPDXX–NH$_2$ | 4.4 |
|  | Ac–DVPAXX–NH$_2$ | >1400 |
| 4th round | Ac–DVPDYX–NH$_2$ | 0.38 |
|  | Ac–DVPDAX–NH$_2$ | >1400 |
| 5th round | Ac–DVPDYA–NH$_2$ | 0.03 |
|  | Ac–DVPDYS–NH$_2$ | 0.27 |
|  | Ac–DVPDYC–NH$_2$ | 0.90 |
|  | Ac–DVPDYV–NH$_2$ | 1.10 |

peptide sequence is represented within the library. A more thorough analysis of the statistical aspects of libraries prepared on beads has been published (Burgess *et al.* 1994). Other groups have calculated the number of beads that have to be used to ensure that all compounds will be present within a defined percentage of the desired proportion (Zhao *et al.* 1995, 1996).

The soluble acceptor molecules (the protein streptavidin and a monoclonal antibody against $\beta$-endorphin) were coupled to either alkaline phosphatase or fluorescein and added to the peptide bead library. Those beads containing peptide sequences that bound to the receptor molecules could be identified by a colour change on the bead, either from the fluorescein or by the phosphatase catalysed modification of a dye molecule. In either case, intensely coloured beads were readily revealed by visual inspection. Micromanipulation with tiny forceps isolated the coloured beads, and following the removal of the receptor molecule by washing, the active peptide sequences were determined by a peptide microsequencing instrument. This technique was highly efficient as several million individual beads could be screened in one afternoon in 10–15 Petri dishes (see Fig. 2.10). Furthermore, the libraries could be washed free of receptor after screening and retained for future testing against other biological targets.

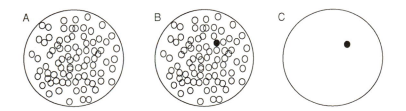

**Fig. 2.10**  (A) The resin beads are arranged in Petri dishes for screening against a soluble receptor molecule. (B) Active compounds are identified by coloration of the bead. (C) Isolation of the coloured bead allows determination of the attached peptide sequence.

**Table 2.3** The peptide sequences with affinity for the monoclonal antibody against $\beta$-endorphin picked from the library

|  | Ligand sequence | $K_i$ (nM) |
| --- | --- | --- |
| Native ligand | YGGFL | 17.5 |
| New ligands | YGGFQ | 15.0 |
|  | YGGFA | 32.9 |
|  | YGGFT | 36.9 |
|  | YGGLS | 726 |
|  | YGALQ | 1980 |
|  | YGGMQ | 8780 |

The monoclonal antibody against $\beta$-endorphin has high affinity for the peptide sequence, Tyr–Gly–Gly–Phe–Leu (YGGFL). A total of six reactive beads were obtained from a total library size of around two million beads. The sequences and the affinities of the discovered ligands are listed in Table 2.3.

Screening the peptide library whilst still attached to the solid phase is a rapid and highly effective way of discovering new peptide ligands for biological targets. But are there limitations to this technique? Unfortunately there are. First, the screening of resin-bound compounds, although an effective way of isolating active compounds, requires the biological target molecule to be in solution, a state available to antibodies and many enzymes, but generally a very difficult requirement for biological receptors which generally only exist in their natural activated state when bound to cell membranes.

Secondly, there are concerns about the validity of testing compounds whilst attached to the solid phase. Medicinal chemists are very aware of the dramatic changes in biological activity that can be effected by the addition of a simple methyl group. How much greater must be the effect of attaching a 100 $\mu$m diameter resin bead. In general, compounds attached to solid phases may be tested if the attachment is on part of the molecule that is usually uninvolved in binding and faces away from the receptor protein into the solvent. Unfortunately, if the chemist is involved in an empirical search for novel activity it is not possible to know which parts of the molecule are necessary for high affinity binding. Peptides are less affected by this problem, as libraries are synthesized with long spacer molecules between the solid phase and the library sequence. However, there is evidence to suggest that the nature of the spacer can dramatically affect binding to target proteins. Work by Yu and Chu (1997) suggests that some spacer molecules can totally abolish biological activity of known active peptide ligands.

Yu and Chu synthesized an epitope (Ac–EGVQQEGAQQPA) from *Borrelia burgdorferi* flagellin, a Lyme disease antigen, on TentaGel resin using either a Pro–Pro, Gly–Gly, or $\beta$-Ala–$\beta$-Ala spacer. The assay was configured such that the beads would turn a deep purple colour if there was binding to H9724, a mouse monoclonal antibody generated against the whole flagellin protein. To their surprise, whilst the beads containing the $\beta$-Ala–$\beta$-Ala spacer showed a dark purple colour, the Pro–Pro linked beads turned only a pale purple. The Gly-Gly spacer conferred a colour midway between

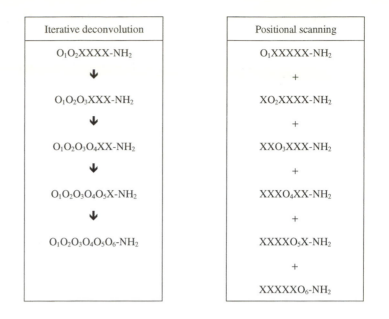

**Fig. 2.11** Comparing iterative deconvolution of mix and split libraries with positional scanning. In each case, 'O' is a residue that is known within the mixture as one of the monomers used to synthesize the library, whereas 'X' is an equimolar mixture of all monomers used. Iterative deconvolution (the so-called '4X' method) requires a sequence of libraries to be made to reveal the final active structure, whereas positional scanning dispenses with the iterative resynthesis, but requires more chemistry to be done 'up-front'.

these two extremes. Clearly the nature of the spacer was having a profound effect on the ability of the ligand to bind to the antibody.

## 6.   Positional scanning

Mix and split synthesis of peptide libraries is an effective way of discovering potent peptide ligands, but the inherent disadvantage of having to deconvolute active mixtures by resynthesis and screening has initiated the development of other ways of making and testing libraries. One alternative is the use of positional scanning libraries, a development first introduced by Houghten (Pinilla *et al.* 1992; Dooley and Houghten 1993). In this technique, rather than make a library just once, several libraries are actually constructed, each comprised of the same mixtures of compounds in different combinations.

One of Houghten's first positional scanning examples concerned a library of 34 million hexapeptide combinations constructed from 18 of the 20 naturally occurring amino acids ($18^6 = 34\ 012\ 224$ sequences). In this library, the same compounds were prepared six times, and in each library a different residue in the sequence is held constant (see Fig. 2.11). Each library consisted of 18 mixtures containing $1\ 889\ 568$ ($18^5$) peptides. The first set of mixtures had the N-terminal amino acid held constant within a mixture (O in Fig. 2.11) but included a random mixture of the 18 amino acids in each of the other positions (X in Fig. 2.11). The identity of the fixed residue in each mixture is known as a consequence of the synthetic approach to the library and thus the most active mixture following screening of this set will define the preferred

N-terminal residue. The second residue is held constant in the second library, and a preferred second residue can be identified from screening these mixtures. The remaining four libraries were constructed using the same approach. The mixtures in each of the six libraries were tested in solution against $\mu$-specific opioid receptors, and the preferred amino acids at each of the six positions were thus revealed.

Two issues raised by positional scanning need to be addressed. First, apart from the final amino acid to be coupled to the growing compounds, the mix and split library paradigm will not easily permit the synthesis of mixtures where other positions can be held constant. Houghten overcame this problem by reacting the resin-bound peptide with a *mixture* of activated amino acids. This then presents the possibility that, as some amino acids acylate more readily than others, the outcome could be unequal quantities, or even omissions of some products. This variation in monomer reactivity was resolved by using an isokinetic mixture—varying concentrations of amino acids dictated by their relative reactivities.

The second potential pitfall with positional scanning is a common concern with the screening of all library mixtures. In short, the most biologically active mixture does not necessarily contain the most active library compound. To circumvent this problem in this positional scanning example, Houghten identifies not just one, but up to four amino acids for each position within the hexapeptide (O1 = Y; O2 = G; O3 = G or F; O4 = F; O5 = F, Y, M, or L; O6 = F, Y, or R), and all combinations of these monomers were synthesized as 24 individual peptides (the most active sequence for inhibition of binding to the $\mu$-specific opioid receptor being YGGFMY, $IC_{50}$ = 17 nM). The suspicion about the most active mixtures not containing the most active compound was borne out by the observation that the hexapeptide based on the most preferred single residue in each position, YGFFFF was only weakly active, possessing an $IC_{50}$ value of 42 $\mu$M.

The theoretical study by workers from Isis (Freier *et al.* 1995), used RNA hybridization of a library of 262 144 9-mers against various short oligomeric targets to examine the success of different library methods. The study concluded that positional scanning is as effective as iterative deconvolution for libraries binding to nine and 18 residue oligonucleotide targets. But for a hexamer target, a compound that bound 26 times weaker than the optimum structure was selected. The failure of positional scanning for this shorter target was attributed to the availability of multiple binding alignments, each with similar binding energies.

It could be argued that such a multiplicity of binding options, although a common observation with oligonucleotide hybridization, is less of a concern with peptides binding to macromolecule targets. However, in a later example of positional scanning, again from Houghten's group (Pinilla *et al.* 1994), evidence for multiple binding modes was observed. A library of no less than four trillion decapeptides were produced and these were screened for the ability to inhibit binding of the monoclonal antibody (mAb) 17/9 to an antigenic peptide (Ac–YPYDVPDYASLRS–NH$_2$). The total library was produced ten times: each set comprised of 20 mixtures that maintained one constant position in the decapeptide.

However it was known that one particular six residue sequence is antigenic for the mAb, and it was expected that this epitope might appear in various

**Table 2.4a** Anticipated preferred amino acids of the mAb 17/9 antigenic sequence, DVPDYA, at each position in the decapeptide library

| | Position | | | | | | | | | |
|---|---|---|---|---|---|---|---|---|---|---|
| | **1** | **2** | **3** | **4** | **5** | **6** | **7** | **8** | **9** | **10** |
| Amino acid | D | V | P | D | Y | A | | | | |
| | | D | V | P | D | Y | A | | | |
| | | | D | V | P | D | Y | A | | |
| | | | | D | V | P | D | Y | A | |
| | | | | | D | V | P | D | Y | A |

**Table 2.4b** Preferred residues actually found in each position following screening

| | Position | | | | | | | | | |
|---|---|---|---|---|---|---|---|---|---|---|
| | **1** | **2** | **3** | **4** | **5** | **6** | **7** | **8** | **9** | **10** |
| Amino acid | D | D | D | D | D | D | A | A | A | A |
| | | | | V | | V | D | D | D | |
| | | | | | | | E | Y | Y | |
| | | | | | | | P | | | |
| | | | | | | | Y | | | |

positions within the decapeptides, leading to a specific distribution of preferred amino acids (Table 2.4a). Indeed this antigenic epitope appeared to 'walk through' the decapeptide sequences, with it being preferentially recognized at the C-terminal region of the peptides (see Table 2.4b).

Thus, having determined that this epitope appeared preferentially at the C-terminus, two new decapeptide libraries were prepared (*N*-acetylated and non-acetylated), and the hexapeptide sequence, DVPDYA, was walked through each. Of the ten possible mixtures, the sequences, Ac–XXXXDVPDYA–NH$_2$ and XXXXDVPDYA–NH$_2$ were found to be the most effective for binding to the mAb (Table 2.5), confirming the results of the original positional scanning library.

Positional scanning can also be applied to cyclic peptides. A library of 82 944 cyclic pentapeptides were made in the search for a known endothelin antagonist (Spatola *et al.* 1996). D-Aspartic acid, a known constituent of the targeted antagonist, the Banyu compound, BQ-123, was used as the tethering point to hydroxymethyl cross-linked polystyrene resin beads, and the library components were built from Boc-protected amino acids, and cyclized on the resin before HF catalysed cleavage. The amino acids employed were the D- and L-forms of six amino acids chosen to provide diverse steric and electronic characteristics as well as to ensure that the appropriate monomers were present to produce the desired compound. The library was constructed with each synthetic step using either a mixture of the 12 protected amino acids

**Table 2.5** The activities of peptide sequences containing the epitope DVPDYA against the mAb 179

| Peptide sequence | $IC_{50}$ (nM) |
| --- | --- |
| Ac–DVPDYAXXXX–NH$_2$ | 8.2 |
| Ac–XDVPDYAXXX–NH$_2$ | 5.5 |
| Ac–XXDVPDYAXX–NH$_2$ | 5.2 |
| Ac–XXXDVPDYAX–NH$_2$ | 6.3 |
| Ac–XXXXDVPDYA–NH$_2$ | 1.6 |
| DVPDYAXXXX–NH$_2$ | 78 |
| XDVPDYAXXX–NH$_2$ | 12.1 |
| XXDVPDYAXX–NH$_2$ | 6.8 |
| XXXDVPDYAX–NH$_2$ | 7.5 |
| XXXXDVPDYA–NH$_2$ | 1.6 |

or a single amino acid, to give mixtures of the format: cyclo(XXXO–D-Asp), cyclo(XXOX–D-Asp), cyclo(XOXX–D-Asp), and cyclo(OXXX-D-Asp). Each of the 48 mixtures thus consisted of $12^3$ or 1728 compounds.

Following the precedent set by Houghten in his pioneering positional scanning work, testing these mixtures against the endothelin receptor generated a range of activities, but reading off the activity from each set suggested that the preferred residues for endothelin receptor affinity were L-Pro at the first position, D-Val at the second, L-Leu at the third, and D-Trp at the fourth. This translates to the structure: cyclo(L-Pro–D-Val–L-Leu–D-Trp–D-Asp) which, reassuringly, is the structure of the known active Banyu endothelin antagonist, BQ-123.

Overall, positional scanning appears to be a highly effective way of screening compounds, but its impact has been attenuated by the sometimes considerable difficulties in making some of the constituent libraries. As mentioned earlier in this section, the difficulty is how to readily introduce the defined residue at any point in the synthetic structure. In 'classical' mix and split, the defined residue is usually the very last residue added to the library. Thus, throughout the synthesis, although at each stage the monomers are added in separate reactions, at the end of each step all the resin is recombined for the next cycle. If it is necessary to define the amino acid at any other position in the synthetic sequence, the mix and split approach needs radical modification. For example, if the residue to be defined is the first to be added to the resin, for every residue in this position there will need to be an individual mix and split procedure. If 20 different amino acids are coupled to the resin in the first round, each of these 20 will be reacted individually to 20 amino acids in the second step. Although these can be combined for the next steps, this is equivalent to increasing the number of reactions 20-fold.

As it is necessary to avoid this amount of synthetic work (which incidentally begins to negate the advantages of 'high speed' combinatorial chemistry) it is feasible instead to react a mixture of protected amino acids with the resin-bound substrates. This alternative approach is achievable with peptide synthesis as the chemistry is highly reliable and high yielding. However, the influence of electronic and steric effects results in different amino acids acylating at different rates. This has been circumvented by the use of mixtures, with the constituent amino acids in a ratio which is determined by their reactivity—the most reactive being present in the lowest relative concentrations. For non-peptide chemistry however, the coupling chemistry is much less reliable, and unless the combinatorial chemist is willing to invest time and resource in measuring relative rates of reaction for his particular set of monomers, there is no way of knowing how to adjust concentrations to take account of varying reactivities. One additional problem which has not been addressed in the synthesis of positional scanning libraries, must be the unseen and unmeasurable effect of reacting mixtures with mixtures. It is likely that individual amino acid monomers may couple preferentially with certain residues on the amino termini of the resin-bound substrates. Despite taking account of the varying reactivities of the amino acids, it is not possible to overcome the possibility of unequal ratios of products being formed within the libraries.

Despite these caveats, positional scanning has been a useful expression of combinatorial synthesis although it seems that it can only be a technique useful for peptide libraries. The underlying principle can be likened to finding the intersection points of activity in a multidimensional matrix of compounds. Less ambitious approaches to active compound identification have been successfully used both on solid phase (see Section 7) and in solution phase libraries (see indexed libraries in Chapter 4, Section 3).

## 7.   Orthogonal combinatorial libraries

Another solution to the problem of finding the most active compound in a large peptide library has been offered through a process entitled 'orthoganol libraries' (Déprez *et al.* 1995). In this process, related to positional scanning, the same peptide library is synthesized twice in mixtures, but the design of the library ensures that each compound is in combination with different analogues in each of the two libraries. In this way, through knowing which two mixtures contain the active component unambiguously suggests the structure of the most active compound in the library. This procedure is most effective if there is only one active compound. If the library contains more than one active compound, the number of possible active compounds increases exponentially, but resynthesis of the possible options would still permit identification of the preferred compounds.

Solid phase peptide synthesis was used to prepare 15 625 trimers resulting from the use of 25 amino acids in each of the three positions. To prepare the libraries, the monomers were pre-combined into mixtures of five, and each of the 125 library mixtures consisted of 125 compounds (5 × 5 × 5). Whereas, one library (A) used one combination of groups of five monomers, the second library (B) was prepared ensuring that the groups of five were distinct. The arrangement of monomers was dictated by the layout of a five by five matrix, with library A using mixtures dictated by the columns, and library B using mixtures represented by the rows (Table 2.6). The distribution of monomers was also arranged to maximize the structural diversity within each mixture. This step was taken to avoid the possibility that mixtures demonstrated misleading cumulative activity from a number of structurally similar active compounds. The final library was constructed such that any mixture from library A and any mixture from library B share only one trimeric product.

**Table 2.6** The arrangement of monomer mixtures used for the orthogonal combinatorial library[a]

|      | A1                  | A2    | A3                                            | A4                          | A5                      |
|------|---------------------|-------|-----------------------------------------------|-----------------------------|-------------------------|
| B1   | D-Leu               | D-Pro | D-Ser                                         | (4-nitro)-D-Phe             | D-Isoglutamic acid      |
| B2   | D-Arg               | D-Ile | Gly                                           | D-Thr                       | D-Isoglutamine          |
| B3   | D-Gln               | D-Tyr | D-Val                                         | D-$\varepsilon$-nicotinoyl-Lys | D-His                   |
| B4   | D-Trp               | D-Asp | D-Glu                                         | D-Methionine sulfoxide      | D-Ala                   |
| B5   | Isonipecotic acid   | D-Lys | D-Tetrahydro-isoquinoleic acid (D-Tic)        | D-Asp                       | D-Phe                   |

[a]Library A was prepared from the mixtures of five monomers represented by the columns, whereas library B used the monomer mixtures represented by the five rows.

The library A was screened against a number of different biological assays. When an active mixture was observed, the second library was screened against the same target. Screening against the vasopressin-2 receptor generated a number of active mixtures; four from library A and three from library B. The intersections of these orthogonal libraries suggest 12 potential active structures. So as to narrow down the choice, active mixtures were selected on the basis of dose–response assays. Through this procedure the most active compound was revealed to be D-Tic–D-Tyr–D-(4-NO$_2$)-Phe.OH (IC$_{50}$ value of 63 nM).

An alternative expression of orthoganol libraries has been described by a group from Merck as spatially arrayed mixtures (SpAM) technology (Berk and Chapman 1997). The same strategy of synthesizing a library twice in different mixture combinations has been used for the preparation of a library of 'peptoids'. Peptoids are oligomers of *N*-substituted glycines, first developed by a group at Chiron (see Chapter 6, Section 5) that have given rise to compounds active at a number of pharmacologically significant receptors. The Merck group prepared the library and deliberately included for rediscovery, a known peptoid $\alpha$1a adrenergic receptor ligand.

The 9216 peptoids prepared in this SpAM library were tetrameric of the form A–B–C–D, prepared on TentaGel resin from sets of 8 halocarboxylic acids A, 12 amines B, 8 amines C, and 12 amines D. The library products had the generic structure depicted in Fig. 2.12.

The two orthoganol libraries prepared were of the formats: $O_A$–$O_B$–$X_C$–$X_D$ and $X_A$–$X_B$–$O_C$–$O_D$, where O is a fixed monomer and X represents a mixture of all monomers. The numbers of precursors were carefully chosen such that a robotic synthesizer could set out the 96 mixtures (each of 96 compounds) of each library into the wells of a 96-well microtitre block. Thus the entire SpAM library fitted into two microtitre plates, with the mixtures configured such that activity in one well of each block uniquely defined the structure of just one active compound. The synthesis of the library $X_A$–$X_B$–$O_C$–$O_D$ proceeded in a

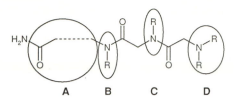

**Fig. 2.12**   The generic structure of compounds in the SpAM library.

straightforward manner, as the mixtures $X_A$–$X_B$ were created on the solid phase using a mix and split approach, and monomers C and D were added singly to the reaction wells. However, for the other library, $O_A$–$O_B$–$X_C$–$X_D$, the first two positions on the resin were the defined monomers. Thus the 96 individual dimers were first prepared, and the mixtures in the C and D positions were introduced through reaction with an isokinetic mixture of monomers.

After synthesis and cleavage from the resin beads, the 192 mixtures were tested against the $\alpha$1a adrenergic receptor. A number of active wells were observed, and from the indexing inherent in this library design, a number of active compounds were identified. Amongst these was the derivative depicted in Fig. 2.13, previously reported to be an $\alpha$1a receptor ligand.

Orthoganol libraries offer several advantages over routine iterative deconvolution in that the evidence of the first round of screening could suggest the structure of the most potent compound in the library. But as the example above demonstrates, if several compounds are active, additional resynthesis and screening is required to define the preferred compound. The requirement for two mixtures to be active to define the most active compound is an internal check on the screening results as it will eliminate problems from false positives. However, care needs to be taken to avoid false negative results as these will result in the real actives being missed. The approach of orthoganol libraries has been applied several times in the synthesis, screening, and deconvolution of solution phase 'indexed' libraries (see Chapter 4, Section 3).

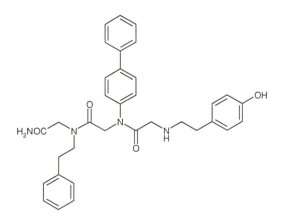

**Fig. 2.13**   The known $\alpha$1a receptor ligand rediscovered from the SpAM library.

## 8. Library nomenclature

This chapter has described a number of different methods for the synthesis and testing of libraries of resin beads. This selection is by no means exhaustive, and although the traditional mix and split remains a favourite, the variety of ways by which combinatorial chemists join monomers to give large and varied libraries is limited by their ingenuity alone. Many other methods using other solid phases are mentioned in later chapters. There have been attempts made to describe how libraries are constructed using a mathematical nomenclature, and amongst these the most comprehensive is by Maehr (1997). Maehr applies mathematical set theory to produce succinct and accurate descriptions of library structure, and then proceeds to use this approach to define methods for the identification of the preformed bioactive component. The mathematical analysis allows a ready assessment of the number of synthetic operations required to prepare the library and also permits a determination of the effectiveness of different library approaches.

This description is complex for the non-mathematical chemist, and whilst it is still too early to say what impact it will have on library design, one lasting legacy of this work may be a general adoption of the accurate terminology Maehr has developed. In Maehr's terminology, a library consists of three fragments A, B, and C derived from monomer sets $A = \{A_1, A_2, ... , A_a\}$, $B = \{B_1, B_2, ... , B_b\}$, and $C = \{C_1, C_2, ... , C_c\}$ such that A, B, and C are represented in $a$, $b$, and $c$ different ways. The term ($^aA\ ^bB\ ^cC$) defines a set of molecules where all the members are present as a mixture—in fact the term ($^aA\ ^bB\ ^cC$) represents the combinatorial product of the three sets of monomers. If the library is prepared in a classical mix and split form through the reactions $A + B \rightarrow AB$ and $AB + C \rightarrow ABC$, the first round synthesis yields the library in a number of partitions $^c(^aA\ ^bB\ C)$ equivalent to the mixtures ($^aA\ ^bB\ C_1$), ($^aA\ ^bB\ C_2$), ($^aA\ ^bB\ C_3$), ..., ($^aA\ ^bB\ C_c$).

The description of this terminology is more straightforward with a real library described by Maehr himself (Maehr and Yang 1997). This example is of a non-peptide library prepared in solution, so formally its description should appear in Chapter 4, but the terms that apply to this library are equally applicable to solid phase and peptide libraries. The monomers A, B, and C correspond to the thiazolyl, aniline, and *N*-acyl portions of a lead leukotriene D4 antagonist, Ro24-5913 (Fig. 2.14) and these were combined to give the combinatorial product of 700 compounds ($^{10}A\ ^7B\ ^{10}C$).

The first round of synthesis generated the C partition $^{10}(^{10}A\ ^7B\ C)$, and these ten mixtures were tested in a LTD$_4$-induced muscle contraction assay

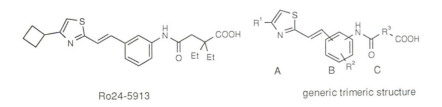

Ro24-5913                    generic trimeric structure

**Fig. 2.14** The structures of Ro24-5913 and the generic trimeric library structure based on the lead compound.

giving a bioactivity data $i_c = \{C_1, C_2, C_3\}$ from which it was concluded that the preferred monomer in this position was $C_1$. The second round of synthesis produced the partition $^7(^{10}A \ B \ C_1)$, testing of which gave the active set $i_b = \{B_1, B_6\}$ and the preferred monomer $B_1$. Finally, testing the partition, $^{10}(A \ B_1 \ C_1)$ produced the active set $i_a = \{A_1, A_3, A_4, A_6, A_7, A_8, A_9\}$. If the active monomers in the sets $i_c$, $i_b$, and $i_a$ are combined into their combinatorial product, this represents 42 compounds that should contain the most potent structures in the library. These 42 were synthesized and assayed, and many active compounds were identified. Of all these however, the most active was the original lead compound, although replacing the cyclobutyl with 4-fluorophenyl or *t*-butyl gave compounds with similar bioactivity.

## 9.  Conclusion

This chapter has focused on the methods of synthesizing combinatorial libraries on resin beads which, after ten years of work, are still the preferred method for solid phase library synthesis. Their popularity derives from the simplicity of performing solid phase chemistry on polymers that has been optimized over the preceding 30 years. The legacy of solid phase peptide synthesis has provided a highly competent springboard for the further development of supported organic chemistry. Library synthesis has also benefited from the advantages of handling and simple purification offered by the resin bead. Techniques to allow rapid screening of libraries either on or off the bead have also aided their wider acceptance by chemists. However, combinatorial chemistry does not end with resin beads. There have been many other solid phases explored for combinatorial synthesis, and these are described in the next chapter.

## References

Arshady, R., Atherton, E., Clive, D.J., and Sheppard, R.C. (1981). Peptide synthesis. Part 1. Preparation and use of polar supports based on poly(dimethylacrylamide). *J. Chem. Soc. Perkin 1*, 529–37.

Atherton, E. and Sheppard, R.C. (1989). *Solid phase peptide synthesis: a practical approach*. IRL Press at Oxford University Press.

Atherton, E., Clive, P.L.J., and Sheppard, R.C. (1975). Polyamide supports for polypeptide synthesis. *J. Am. Chem. Soc.*, **97**, 6584–5.

Atherton, E., Logan, C.J., and Sheppard, R.C. (1981). Peptide synthesis. Part 2. Procedures for solid-phase synthesis using fluorenylmethoxycarbonylamino-acids on polyamide supports. Synthesis of substance P and acyl carrier protein 65–74 decapeptide. *J. Chem. Soc. Perkin 1*, 538–46.

Bayer, E., Dengler, M., and Hemmasi, B. (1985). Peptide synthesis on the new polyoxyethylene-polystyrene graft copolymer, synthesis of insulin B21-30. *Int. J. Pept. Protein Res.*, **25**, 178–86.

Berk, S.C. and Chapman, K.T. (1997). Spatially arrayed mixture (SpAM) technology: synthesis of two-dimensionally indexed orthoganol combinatorial libraries. *Bioorg. Med. Chem. Lett.*, **7**, 837–42.

Burgess, K., Liaw, A.I., and Wang, N. (1994). Combinatorial technologies involving reiterative division/coupling/recombination: Statistical considera-tions. *J. Med. Chem.*, **37**, 2985–7.

Déprez, B., Williard, X., Bourel, L., Coste, H., Hyafil, F., and Tartar A. (1995). Orthogonal combinatorial chemical libraries. *J. Am. Chem. Soc.*, **117**, 5405–6.

Dooley, C.T. and Houghten, R.A. (1993). The use of positional scanning synthetic peptide combinatorial libraries for the rapid determination of opioid receptor ligands. *Life Sci.*, **52**, 1509–17.

Erb, E., Janda, K.D., and Brenner, S. (1994). Recursive deconvolution of combinatorial libraries. *Proc. Natl. Acad. Sci. USA*, **91**, 11422–6.

Freier, S.M., Konings, D.A.M., Wyatt, J.R., and Ecker, D.J. (1995). Deconvolution of combinatorial libraries for drug discovery. *J. Med. Chem.*, **38**, 344–52.

Furka, A., Sebestyen, F., Asgedom, M., and Dibo, G. (1988). *Abstr. 14th Int. Congr. Biochem. Prague, Czechoslovakia*, **5**, 47.

Furka, A., Sebestyen, F., Asgedom, M., and Dibo, G. (1991). General method for rapid synthesis of multicomponent peptide mixtures. *Int. J. Pept. Protein Res.*, **37**, 487–93.

Grubler, G., Stoeva, S., Echner, H., and Voelter, W. (1994). *Peptides: chemistry, structure and biology* (ed. R.A. Hodges and J.A. Smith), pp. 51–3. ESCOM-I11, Leiden.

Houghten, R.A. (1985). General method for the rapid solid-phase synthesis of large numbers of peptides: Specificity of antigen-antibody interaction at the level of individual amino acids. *Proc. Natl. Acad. Sci. USA*, **82**, 5131–5.

Houghten, R.A., Bray, M.K., Degraw, S.T., and Kirby, C.J. (1986). Simplified procedure for carrying out simultaneous multiple hydrogen fluoride cleavages of protected peptide resins. *Int. J. Pept. Protein Res.*, **27**, 673–8.

Houghten, R.A., Pinilla, C., Blondelle, S.E., Appel, J.R., Dooley, C.T., and Cuervo, J.H. (1991). Generation and use of synthetic peptide combinatorial libraries for basic research and drug discovery. *Nature (London)*, **354**, 84–6.

Köster, H., Stumpe, A., and Wolter, A. (1983). Polymer support oligonucleotide synthesis 13: rapid and efficient synthesis of oligonucleotides on porous glass support using triester approach. *Tetrahedron Lett.*, **24**, 747–50.

Lam, K.S., Salmon, S.E., Hersch, E.M., Hruby, V.J., Kazmierski, W.M., and Knapp, R.J. (1991). A new type of synthetic peptide library for identifying ligand-binding activity. *Nature (London)*, **354**, 82–4.

Lam, K.S., Hruby, V.J., Lebl, M., Knapp, R.J., Kazmierski, W.M., Hersch, E.M., *et al.* (1993). The chemical synthesis of large random peptide libraries and their use for the discovery of ligands for macromolecular acceptors. *Bioorg. Med. Chem. Lett.*, **3**, 419–24.

MacDonald, A.A., Dewitt, S.H., Ghosh, S., Hogan, E.M., Kieras, L., Czarnik, A.W., *et al.* (1995). The impact of polystyrene resins in solid-phase organic synthesis. *Mol. Diversity*, **1**, 183–6.

Maehr, H. (1997). Combinatorial chemistry in drug research from a new vantage point. *Bioorg. Med. Chem.*, **5**, 473–91.

Maehr, H. and Yang, R. (1997). Optimization of a leukotriene D4 antagonist by combinatorial chemistry in solution. *Bioorg. Med. Chem.*, **5**, 493–6.

Merrifield, R.B. (1963). Solid phase peptide synthesis. I. The synthesis of a tetrapeptide. *J. Am. Chem. Soc.*, **85**, 2149–54.

Pinilla, C., Appel, J.R., Blanc, P., and Houghten, R.A. (1992). Rapid identification of high affinity peptide ligands using positional scanning synthetic combinatorial libraries. *BioTechniques*, **13**, 901–5.

Pinilla, C., Appel, J.R., and Houghten, R.A. (1994). Investigation of antigen–antigen interactions using a soluble, non-support-bound synthetic decapeptide library composed of four trillion ($4 \times 10^{12}$) sequences. *Biochem. J.*, **301**, 847–53.

Pirrung, M.C. (1993). Encoded combinatorial peptide libraries containing non-natural amino acids. *Chemtracts: Org. Chem.*, **6**, 88–91.

Quarrell, R., Claridge, T.D.W., Weaver, G.W., and Lowe, G. (1996). Structure and properties of TentaGel resin beads: Implications for combinatorial library chemistry. *Mol. Diversity*, **1**, 223–32.

Rapp, W.E. and Bayer, E. (1994). *Peptides: chemistry, structure and biology* (ed. R.S. Hodges and J.A. Smith), pp. 40–3. ESCOM-I11, Leiden.

Rapp, W., Zhang, L., and Bayer, E. (1990). *Innovation and perspectives in solid phase synthesis; 1st Intl. Symposium* (ed. R. Epton), pp. 205–10. SPCC(UK) Ltd. Birmingham.

Scott, L.T., Rebek, J., Ovsyanko, L., and Sims, C.L. (1977). Organic chemistry on the solid-phase. Site-site interactions on functionized polystyrene. *J. Am. Chem. Soc.*, **99**, 625–6.

Spatola, A.F., Crozet, Y., de Wit, D., and Yanagisawa, M. (1996). Rediscovering an endothelin antagonist (BQ-123): A self-deconvoluting cyclic pentapeptide library. *J. Med. Chem.*, **39**, 3842–6.

Stahl, G.L., Smith, C.W., and Walter, R. (1980). Superior swelling properties of resins of poly-*N*-acrylyldialkylamines over polystyrene in solvents for peptide synthesis. *Int. J. Pept. Protein Res.*, **15**, 331–4.

Szymonifka, M.J. and Chapman, K.T. (1995). Magnetically manipulable polymeric supports for solid phase organic synthesis. *Tetrahedron Lett.*, **36**, 1597–600.

Virgilio, A.A. and Ellman, J.A. (1994). Simultaneous solid-phase synthesis of $\beta$-turn mimetics incorporating side-chain functionality. *J. Am. Chem. Soc.*, **116**, 11580–1.

Yu, A. and Chu, Y-H. (1997). Combinatorial epitope search: pitfalls of library design. *Bioorg. Med. Chem. Lett.*, **7**, 95–8.

Zhao, P-L., Zambias, R., Bolognese, J.A., Boulton, D., and Chapman, K. (1995). Sample size determination in combinatorial chemistry. *Proc. Natl. Acad. Sci. USA*, **92**, 10212–16.

Zhao, P-L., Nachbar, R.B., Bolognese, J.A., and Chapman, K. (1996). Two new criteria for choosing sample size in combinatorial chemistry. *J. Med. Chem.*, **39**, 350–2.

# 3    Other solid phases

Unlike the synthesis of compound libraries on resin beads, the use of other solid phases such as multipins and laminar solid phases can be generally described as spatially addressable combinatorial methods. Usually, the compounds made on these supports are discrete, single, and made simultaneously by an efficient combinatorial method. During screening, the library will yield information instantly on individual components, the identity of the compounds being determined by their spatial position within the library array. This is not to say that resin bead-based synthesis cannot be used for a spatially addressable library. In fact these methods of making single compounds in a spatially addressable form using resin samples in discrete tubes or positions in a 96-well microtitre block are gaining in popularity.

## 1.    Libraries on multipins

The polyethylene 96-well microtitre plate is commonly used in immunological studies for the screening of peptides, and is increasingly the mainstay of high-throughput pharmaceutical testing. The 8 × 12 array layout of the wells is convenient for carrying out multiple dilutions and biological assays. It was this structured layout that initiated one of the first steps in the development of combinatorial chemistry. Geysen, working at the Commonwealth Serum Laboratories in Australia developed a significant alternative to the resin bead for solid phase peptide synthesis: the reusable polyacrylic acid grafted polyethylene rod (Geysen *et al.* 1984). These 'multipins', 40 mm in length and with a diameter of 4 mm were gamma irradiated at a dose of one million rads in a 6% aqueous solution of acrylic acid. Under such intense irradiation, active radical species are generated on the surface of the pins. These are subsequently trapped by radical addition to the acrylic acid to generate a surface coating of polyacrylates with carboxylic acid functionality available for further chemical derivatization. The carboxylic acid was derivatized with a diamine-β-alanine–β-alanine spacer, both to increase mobility and reactivity of the terminal functionality and also to provide an amine group that offered a suitable functionality for attaching peptides (Fig. 3.1).

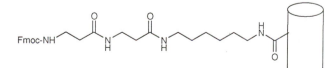

**Fig. 3.1**    The diamine β-alanine-β-alanine linker group attached to the pins for peptide synthesis.

The pins were arranged in the microtitre plate format by attachment to a supporting block, allowing 96 different peptides to be synthesized simultaneously, and also to allow microtitre plates to be used as reservoirs for reaction solutions. The 96 wells could contain different amino acid solutions to permit independent synthesis on each of the pins, but common steps such as washings and deprotections could be carried out simultaneously in a common reaction vessel. In the original studies, the carboxylic acid derivatization was found to be 0.15–0.2 $nmolmm^{-2}$ leading to between 10 and 100 nmol of peptide (or up to $6 \times 10^{16}$ molecules) on each pin. Peptides were synthesized using standard Boc-protection/carbodiimide coupling chemistry and the products could either be cleaved into a microtitre plate or tested whilst still attached to the pin. In the latter case the pins could be reused up to 60 times for a variety of targets including interactions with monoclonal antibodies or polyclonal sera.

Geysen set himself the challenge of using the pins for epitope mapping, the process of making and testing every short peptide along the complete sequence of an immunologically active protein, searching for the short amino acid sequence responsible for binding to the antibody. Although the numbers of peptides required to be synthesized are not great by today's standards, at the time the prospect of synthesizing several hundred peptides for such a study in a realistic time-frame was an almost impossible demand. Multipins for the first time offered the level of synthetic productivity required to meet this need.

The initial biological study was the discovery of the immunogenic peptide of the foot and mouth disease virus (FMDV) (Geysen *et al.* 1984, 1985), possibly one of the most significant successes of epitope scanning. It was demonstrated that the immunogenic epitope of the coat protein (VP1) of FMDV, type $O_1$, was located at residues 146–152 of that protein. This result was achieved by the synthesis of all 208 possible overlapping hexapeptides covering the 213 residue length of the coat protein. The epitope sequence was suggested by the observation that hexapeptides 146 (GDLQVL) and 147 (DLQVLA) demonstrated affinity for antisera. However, at this stage it was not possible to say which amino acids common to these two hexapeptide sequences constituted the critical portion of the immunogenic epitope. Thus in a separate study, each of the amino acids in the peptide 146 were exchanged in turn with all 20 natural amino acids, giving a total of 120 different hexapeptides all prepared on multipins. It was found that antibody binding was most sensitive to changing the two leucine residues at positions 148 and 151, and thus these amino acids were deemed essential for recognition by the antisera raised against the intact virus.

Geysen had thus found a very neat solution to the requirements of epitope mapping. He had found that generally, the peptide epitope sequence that is recognized by an antibody is up to eight amino acids long. Average epitope length is around six residues of which five are considered essential for binding to the antibody. Epitope mapping on multipins works well provided that the epitope is a contiguous sequence within the antigenic protein; Geysen's approach was successful because he prepared all of the hexapeptides within the immunogenic protein. However, some epitopes are represented by amino acids that may be in close proximity in the folded protein but are not contiguous within the primary sequence. If this is the case, it is almost impossible to find the epitope through the mapping approach. Instead it is necessary to make every possible hexapeptide sequence of naturally occurring amino acids. It is clearly no longer feasible to prepare such a large library as single components on pins, as the entire library would require 64 million individual pins—a feat beyond even the most hard-working peptide chemist. Houghten solved this problem by the synthesis of peptide libraries in mixtures using 'tea-bags' containing resin beads (see Chapter 2, Section 3).

Geysen's solution was to generate a *mixture* of peptides on each pin so that the entire library of peptides was contained on many fewer than 64 million pins. Each pin initially contained a mixture of hexapeptides of the format XXOOXX, where X represents the presence of all 20 amino acids and O represents a single defined residue. Using this approach, all 64 million peptides could be represented on just 400 pins, each containing a mixture of 160 000 peptides, each sequence represented by around $4 \times 10^{11}$ molecules. This approach was applied in the search for the epitope that binds to an antibody raised against the sequence DFLEKI of the protein myohemerythrin (Geysen and Mason 1993). Following the first round of screening, the pin that generated the greatest antibody affinity led to the identification of the central two amino acid residues (L and E). The next round of synthesis prepared pins containing peptides with the sequences, XOLEXX and XXLEOX, exploring the residues either side of the two defined amino acids. Following a procedure of iterative resynthesis and screening, a deconvolution not at all dissimilar to the process used in mix and split bead libraries, the sequence DFLEKI was rediscovered.

Although on this occasion a contiguous epitope was identified, the process of preparing every possible hexapeptide sequence allows the possibility of discovering discontinuous epitopes. Alternatively it may be possible to find peptide sequences that mimic the chemical properties of the epitope, so-called mimotopes, but that bear no direct relationship to the primary sequence of the immunogenic protein. Geysen makes the point that at each stage of the deconvolution, the scientist may be faced with several mixtures of roughly equivalent potency. The choice then needs to be taken as to which residue or residues are followed up for deconvolution in the next round of synthesis. This decision is not trivial, as the expectation that all solutions will converge to a preferred epitope sequence is rarely the case. Each of the amino acid positions is not independent, and the choice of a particular residue in one position will influence the preferred residues in other positions in the peptide sequence. Thus, whilst one series of choices may lead to an epitope that is identical to the contiguous sequence on the immunogenic protein, any deviation from this path will lead to a mimotope and probably not the globally optimum sequence.

The multipins solid phase presented a solution to a problem that had limited immunology studies for many years. Although the pioneering work of Merrifield had provided efficient methods for the synthesis of peptides, cost and resource limitations prevented the comprehensive synthesis of large numbers of peptides to investigate the sites and nature of protein epitopes. Consequently, the choice of which peptides to make had become the subject of predictive algorithms. The multipin approach, by relaxing criteria for quantity and purity, allowed the numbers of peptides that could be readily synthesized to be increased a 1000-fold. A systematic approach by the synthesis of many peptides could provide a comprehensive structure–activity database that would more rapidly answer key questions of epitope structure. A range of epitope identification techniques possible using peptides synthesized and assayed on grafted pins have been listed by Geysen (Geysen *et al.* 1987).

Subsequently, synthesis on multipins was modified to allow release of the peptides for solution studies (Bray *et al.* 1990, 1991*a*). This was achieved by the initial synthesis of a Boc.Lys(Fmoc)–Pro dipeptide on the linker attached to the solid phase. The peptide sequence was built on the ε-side-chain amine of lysine using Fmoc protection, and at the end of the synthesis, the complete peptide could be released by TFA catalysed deprotection of the Boc group, and subsequent diketopiperazine formation (see Fig. 3.2). Although the cleavage resulted in all product peptides containing a diketopiperazine residue at the C-terminus, this was claimed not to interfere with biological assays.

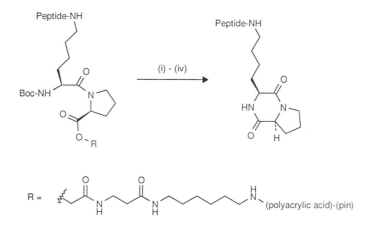

**Fig. 3.2**   Peptide cleavage into pH 7 buffer via diketopiperazine formation. Conditions: (i) TFA/phenol/ethanedithiol; (ii) sonication; (iii) pH 3 buffer; (iv) pH 7 buffer.

To obviate any possible problem conferred by the presence of the diketopiperazine, conditions were later developed to allow peptide release as C-terminal carboxamides by exposing the pins to ammonia vapour and eluting the peptides off with aqueous ethanol (Bray *et al.* 1991*b*) and this approach has been used for a study of substance P analogues (Wang *et al.* 1993). Peptide amides have also been prepared by incubating the pins with ammonia/ tetrahydrofuran vapour (Bray *et al.* 1994). The co-solvent was shown to be essential to achieve sufficient solvation of the solid phase, although the choice of THF was critical as the use of ammonia/methanol led to product peptides contaminated with the methyl ester. Following ammonolysis the cleaved carboxamides could be washed from the solid support with a solvent of choice.

The pin has recently been redesigned (Valerio *et al.* 1993) to take a detachable four pronged polyethylene crown: an injection moulding 5.5 mm high, 5.3 mm wide, with a total surface area of 130 mm². The crown was irradiated and derivatized with 2-hydroxyethyl methacrylate (HEMA) monomer giving a total functional group loading in the range 1.0–2.2 μmol per crown. HEMA was found to give the most reproducible grafting, with consistently higher loadings than were attainable with acrylic acid and furthermore, the hydroxy group allowed the direct addition of amino acids through an ester linkage. In fact the HEMA linker has been used with the addition of 4-hydroxymethylphenoxyacetic acid for the generation of carboxylic acids, and with *p*-[α-(Fmoc-amino)methyl]-2,4-dimethoxybenzyl-phenoxyacetic acid for the generation of carboxamides (Fig. 3.3) (Valerio

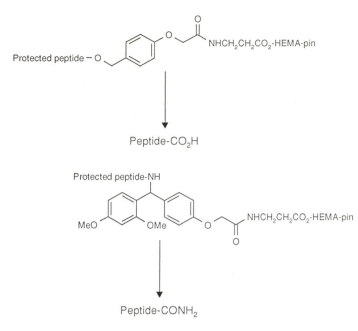

Peptide-CO$_2$H

Peptide-CONH$_2$

**Fig. 3.3** Two groups attached to the grafted pins through HEMA linkers that yield carboxylic acids and carboxamides following treatment with TFA/scavenger mixtures.

*et al.* 1994). In each case the final peptide was isolated by treatment with a TFA/scavenger mixture, and washed from the pin with a series of ether/petroleum wash and centrifugation steps.

The multipin method is an effective method for the parallel synthesis of peptides as both carboxylic acids and carboxamides, but with the increasing interest in combinatorial libraries of non-peptide structures, this technology, as with many other originally designed for peptide use, has been modified to make non-peptides. For example, a set of 11 β-turn mimics were simultaneously synthesized on multipins (Virgilio and Ellman 1994) (Fig. 3.4). β-Turns are one of the three major secondary structural elements in peptides and proteins, and thus play a key role in molecular recognition between peptides and proteins. Compounds that mimic these recognition elements have potential as drug molecules and are seen as attractive targets for the synthesis of lead discovery combinatorial libraries. Ellman's group (Bunin *et al.* 1994) has also prepared a library of 192 structurally diverse 1,4-benzodiazepines on two 12 by 8 arrays of grafted pins using the solid phase synthetic route described in Chapter 7, Section 2. One necessary modification of the pins technology was the need to perform air-sensitive reactions. This was achieved by washing the pins in dry THF prior to placing the pins and all necessary reagents in a glove bag flushed with dry nitrogen gas. The pins were immersed in 0.12 M lithiated 4-phenylmethyl-2-oxazolidinone in THF at 0 °C, to generate the amide anion, and then immersed without washing into preformed solutions of the alkylating agents in DMF (Fig. 3.5). To increase yields, the alkylation procedure was repeated before allowing the pins to warm to ambient temperature. After cleavage from the solid support, the benzodiazepines were shown to have been produced in an average chemical yield of 86%.

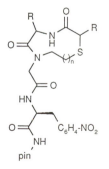

**Fig. 3.4** The structure of β-turn mimics synthesized on multipins.

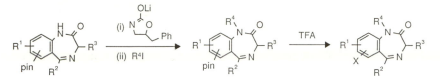

**Fig. 3.5**   Ellman's solid phase synthesis of benzodiazepines on multipins.

Multipin solid supports have also been used for the library synthesis of 4-aminoproline analogues (Bray *et al.* 1995), acyl 2,3-diaminopropionic acid oligomers (Valerio *et al.* 1996*a*), and ethers through the Mitsunobu reaction (Valerio *et al.* 1996*b*). In both the first and third of these papers, parallel synthesis of multipins was used to optimize the synthetic conditions. To find the best yields for the reductive amination of a solid-supported ketoproline derivative, five different reaction parameters: pH, type of amine, concentration of amine, concentration of reducing agent, and alcohol solvent were varied simultaneously (Fig. 3.6). The products were cleaved from the pins and analysed by high-throughput MS and HPLC. It was found that high yields were favoured by higher amine concentrations and lower pH and that methanol was generally the preferred solvent.

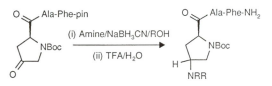

**Fig. 3.6**   The reductive amination of 4-ketoproline performed on multipin solid support.

Not every type of compound can be readily prepared on multipins. Attempts to prepare 1,4-benzodiazepine-2,5-diones were frustrated by the unavailability of a suitable chloromethyl functionality. Despite engineering a suitable functionality, variable yields in the synthesis were observed, indicating a requirement for substantial chemistry optimization prior to library synthesis (Boojamra *et al.* 1997). A library of benzodiazepinediones was subsequently successfully prepared on Merrifield resin.

Multipins were originally developed as a method for accelerating the synthesis of peptides for immunological studies. Like many other rapid synthesis methods, the technique has been 'highjacked' by combinatorial chemistry, with a steady but significant move towards higher loadings, more functionally varied solid phase materials, and ultimately more diverse chemistry. Multipins have been used for analogue synthesis and reaction optimization, but it is in the synthesis of peptides for epitope studies that multipins have had their greatest impact and they have made their mark as one of the most valuable combinatorial chemistry techniques.

## 2.   Libraries on winks

Disks of porous polyethylene have been used for the synthesis of peptide positional scanning libraries (Luo *et al.* 1995). These 'winks' (presumably named after 'tiddlywinks' due to their size and shape), a quarter inch in diameter and one-eighth of an inch thick, were oxidized with chromic acid to functionalize the surface with carboxylic acid groups. Then through a multistep derivatization sequence, a diamine spacer and a carboxymethylated dextran was linked. Subsequent addition of a diamine to the carboxylic acid on the dextran provided the handle for peptide synthesis. Overall, this complex and lengthy linker (Fig. 3.7) was designed both to provide a highly hydrophilic environment to permit peptide synthesis and to allow unhindered access to solution proteins for biological assay.

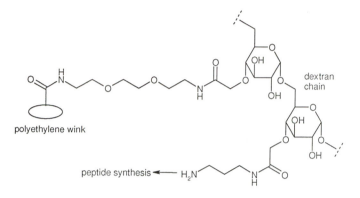

**Fig. 3.7** The linker attached to polyethylene winks for the attachment of peptide libraries.

A series of libraries were synthesized as mixtures on the individual winks constituting in total a positional scanning library (see Houghten's work described in Chapter 2, Section 6). This library was designed to investigate the peptide sequence specificity of a number of kinases. Kinases are enzymes that work by phosphorylating the hydroxyls of specific serine, tyrosine, or threonine residues, and these libraries were used firstly to investigate cyclic adenosine monophosphate (cAMP)-dependent protein kinase A (PKA). The library mixtures (Table 3.1) were based on the sequence of kemptide (LRRASLG), a known PKA peptide substrate.

**Table 3.1** The peptide libraries synthesized in the study to explore peptide sequence specificity of protein kinases[a]

| Library | −6 | −5 | −4 | −3 | −2 | −1 | 0 | +1 | +2 | +3 |
|---|---|---|---|---|---|---|---|---|---|---|
| X − 3 library | | | L | X | O | O | S | L | G | |
| X − 2 (R − 3) library | | | L | R | X | O | S | L | G | |
| X − 2 (K − 3) library | | | L | K | X | O | S/T | O | O | O |
| X − 1 library | | | L | K | O | X | S/T | O | O | O |
| X + 1 library | | | L | K | O | O | S/T | X | O | O |
| X + 2 library | | | L | K | O | O | S/T | O | X | O |
| X + 3 library | | | L | K | O | O | S/T | O | O | X |
| X − 4 library | O | O | X | K | O | O | S/T | L | G | |
| X − 5 library | O | X | O | K | O | O | S/T | L | G | |
| X − 6 library | X | O | O | K | O | O | S/T | L | G | |

[a]O = mixture of 17 amino acids (not C, S, or T) and X = one of 19 amino acids (not C). The amino acids are numbered relative to the phosphorylated residue at position zero.

Testing the peptide mixtures whilst attached to the wink and identifying the preferred residues in each of the positions marked by X, it was found that the preferred substrate for PKA is RRXS(I/L/V). Referring to the phosphorylated serine as position 0, there is no preference for the X − 1 position, and X + 1 is preferably I, L, or V, one of three moderately lipophilic aliphatic amino acids.

The same wink-based libraries were also used to determine the sequence preference of two transforming growth factor $\beta$ (TGF-$\beta$) receptors (T$\beta$RI and T$\beta$RII). The optimal peptide sequences were found to be KKK(K/R)KK (S/T)XXX for T$\beta$RI, and a very similar sequence, K(K/R)KKKK(S/T)XXX for T$\beta$RII.

The solid phase wink is a specialized support which plays a similar role in this study to the resin beads in Houghten's positional scanning peptide library, or a mixture library on Geysen's multipins. Although the authors refer to winks as a spatially addressable library, it is such only inasmuch as the identity of the mixtures on the individual winks is known from their locations within a set of labelled reaction vessels.

## 3.   Libraries on laminar solid phases

For centuries, chemistry has been carried out in single vessels, usually made from glass, and whilst the chemist has his sights on the production of a single compound, this technology has been perfectly adequate. With the demands of biological screening the need to run many reactions in parallel has led to the design and development of multiplexed reaction vessels. For much of this work, chemists have borrowed the microtitre block as it offers 96 reaction wells moulded into the same vessel. The block can be visualized as a two-dimensional planar matrix of separate reactions. To overcome the inflexibility of the 96-well plate, chemists are exploring other planar arrays of separate reactions, and one extension of this concept is the development of a planar solid phase used in a technology that can address specific sites for individual reactions.

This section explores the several different approaches to this goal—the development of a planar solid phase for multiple chemical reactions.

### 3.1   Light-directed parallel synthesis

During the early years of the 1990s when the first disclosures of combinatorial chemistry were being made, one technique in particular elicited much interest, primarily because it represented a marriage between chemistry and semiconductor design and manufacture. Such a combination was perhaps not surprising from a group at Affymax, a company located next to Silicon Valley in California. This group published a seminal paper in the journal *Science* describing the technology of VLSIPS—very large scale immobilized peptide synthesis (Fodor *et al.* 1991). This name was chosen to make an implicit recognition of the parallel with very large scale integrated (VLSI) circuit technology exploited in the semiconductor industry. One of the key inventors has written a highly readable account of the development of this technology (Pirrung 1997).

The method allowed the synthesis of large numbers of peptides or oligonucleotides on a glass slide, with each unique sequence occupying a specific region of space. Because the library was screened whilst still attached to the support, biologically active compounds, detected by a fluorescent reporter group could be identified by their location. The synthesis was so precise that in one of the first reported syntheses, each compound occupied a region on the glass support just 50 $\mu$m × 50 $\mu$m.

To prepare the library, orthodox peptide coupling methods were used combined with a protection strategy that employed light-labile protecting groups. Photolithography, the key to the revolution in microelectronics, was the technique used to allow such precise placing of the library compounds. The amino acids employed were protected on the nitrogen with the 6-nitroveratryloxycarbonyl (NVOC) group, that could be cleaved by photolysis by 365 nm UV light (Fig. 3.8).

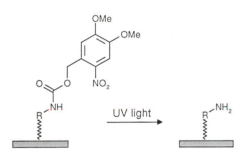

**Fig. 3.8** The 6-nitroveratryloxycarbonyl (NVOC) protecting group is cleaved by UV irradiation to generate free amino groups for further derivatization.

The synthesis was performed on a microscope glass surface that was firstly derivatized with 3-(aminopropyl)triethoxysilane, followed by reaction with aminohexanoic acid. The free amino groups were finally protected with the NVOC group. Deprotection of specific sites on the solid support was possible by the use of photolithography masks. If only half of the support is covered by the mask and the glass irradiated with UV light, the chemical functionality on the exposed half will be deprotected (Fig. 3.9). The glass slide is now subjected to coupling conditions to add the first amino acid (A) by total immersion in a chemical reactor containing the hydroxybenzotriazole (HOBt) ester of the amino acid. In this coupling reaction only the functionality on the half of the support that has been deprotected is available for derivatization.

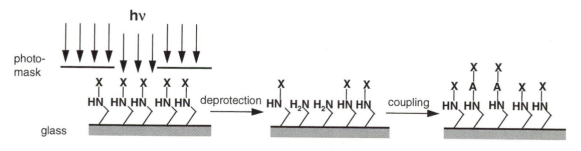

**Fig. 3.9** The use of a photomask allows specified regions of the glass chip to be irradiated leading to deprotection and subsequent derivatization of the chemical functionality in those regions.

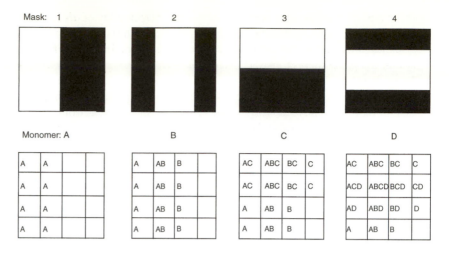

**Fig. 3.10** A sequence of photolithography masks that can be used to synthesize all 16 monomer permutations of a specific tetrameric sequence.

Half of the slide will be derivatized with the amino acid and the other half still bears only the protected amine functionality. If a second mask with a translation relative to the first is applied and the slide irradiated, the two zones on the slide will be subdivided into four by deprotection of half of the residues on the support. Addition of a second amino acid (B) will generate a different species in each quarter of the support (AB, A, B and nothing). Using further masks to deprotect specific regions of the glass support and further coupling with monomers C and D leads to the production of 16 unique regions on the glass support each bearing a different peptide sequences. Figure 3.10 illustrates this technique and it can be observed that one-sixteenth portion of the glass slide contains the full tetrapeptide, another sixteenth contains nothing, and the other 14 regions contain every permutation of compound in between. By using further photomasks and adding different amino acids permits longer and more complex combinations of peptide to be assembled. In every case, the pattern and sequence of masks used defines the nature and the location of the library products.

The above binary masking strategy generates $2^n$ compounds in $n$ chemical steps. There is an alternative masking strategy using striped masks that will generate $n^l$ different compounds where there are $n$ building blocks in each of the $l$ positions of the oligomer. Unlike the binary approach, there is no necessity to include deletion products within the library, and all compounds synthesized will have the same length.

This binary procedure was extended to ten steps in total using a predetermined series of opaque masks, to generate a total set of 1024 peptide sequences in a 32 × 32 array generating reaction sites 400 $\mu$m × 400 $\mu$m. As would be anticipated, one particular position on the glass support was masked for every deprotection and thus did not have a peptide attached and this acted as a negative control in subsequent biological assays. There were ten positions with a single amino acid, 45 with dipeptides, and then 120, 210, 252, 210, 120, 45 and one positions with three, four, five, six, seven, eight, nine and ten residues respectively.

The library was screened whilst still attached to the solid phase against the fluorescein-labelled 3E7 mAb raised against $\beta$-endorphin, and sequences that bound were detected by a fluorescence scan, and identified by their location within the 32 × 32 matrix. The monomers employed for the synthesis, in chronological sequence were F, S, L, F, T, G, A, G, Y, and F, and the 15 most active peptides discovered were YGAFLS, YGAFS, YGAFL, YGGFLS, YGAF, YGALS, YGGFS, YGAL, YGAFLF, YGAF, YGAFF, YGGLS, YGGFL, YGAFSF, and YGFLSF. An obvious feature in these active peptides is that they all have Tyr–Gly at the N-terminus which is in agreement with previous observations that an N-terminal tyrosine is a key determinant of binding to this antibody.

The technique of VLSIPS is clearly a very powerful extension of the combinatorial synthesis of peptides. It consumes very small quantities of reagents and monomers and can be used to make huge numbers of compounds on very small areas of solid support. However, it suffers from the limitations imposed by screening on the solid phase, as this is essential to keep the information on library product locations and thus permit the identification of active sequences. In the years following the first disclosure of the VLSIPS technology, it has not been widely applied for the production of combinatorial libraries. However it is now finding a niche and having a significant impact in DNA identification and sequencing. The explanation for this can be gleaned from an examination of the mathematics of combinatorial library production. The number of different oligomeric sequences possible is an exponential function of the length of the oligomer. However, the number of synthetic steps required to generate the library is the product of the length of the oligomer and the number of monomers used. For example, a tetrapeptide library will require 80 steps to produce 160 000 ($20^4$) different sequences. By contrast, 80 synthetic steps will generate $4^{20}$ or almost 1.1 million million different oligonucleotide sequences, and is thus a much more efficient diversity generator. As a library of all possible 16 base oligomers will contain more sequences (4.3 billion) than occur in the human genome, such a library would be a very powerful DNA sequencing tool and would be achievable in very few synthetic steps.

The concept of preparing oligonucleotide arrays on glass slides has also originated in the laboratory of Maskos and Southern (1992a). Their approach employs reaction 'channels' formed from lines 1–10 mm apart of 1.2 mm silicon rubber tubing. This tubing acts as a gasket when pressed against the glass reaction surface and solutions of phosphoramidites are introduced by capillary action along the channels. The linker used for the synthesis was designed specially for this work and depended on the reaction of the silanol glass surface with 3-glycidoxypropyltrimethoxysilane (Maskos and Southern 1992b) (Fig. 3.11). When using wet solvent or carrying out the reaction in water, the side-chains underwent further cross-linking. In the second step the epoxide could be cleaved with water or a diol to provide a free hydroxyl group for the synthesis of oligonucleotides.

By applying a number of different physical masks it was possible to subdivide the glass slide into a number of regions containing different oligonucleotide products. Between the coupling steps, the entire glass slide could be reacted in batch mode with dichloroacetic acid to deprotect the 5′-protecting group. Using this approach, four different oligonucleotide sequences of the form $G_3A_3$, $G_6$, $A_6$, and $A_3G_3$ were synthesized in parallel lines on one glass slide, and other glass slides contained longer sequences of

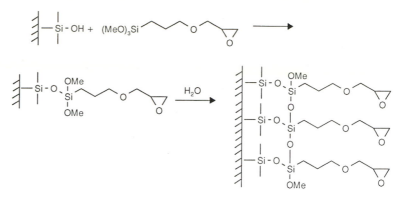

**Fig. 3.11**   The synthesis of linkers on glass surface for oligonucleotide synthesis.

the same base patterns. Hybridization studies with probe oligonucleotides were complete in 30–120 minutes.

A second generation method has also been developed in the same laboratory (Southern *et al.* 1994). In this approach, the striped reaction channels were abandoned in favour of a regular geometrically-shaped reaction zone that was moved linearly across the glass reaction surface. After each coupling step the mask was moved sufficiently to cover new underivatized glass, as well as overlapping with previously functionalized surface. Figure 3.12 illustrates the effect for a diamond-shaped mask, and it can be seen that some regions of the glass surface (at the edge) will carry only one nucleotide, but others (towards the centre) will carry oligonucleotides of varying lengths covering all the contiguous 'scanning' oligomers of a particular predetermined sequence. The maximum length of oligonucleotide sequence will be defined by the size of the masking zone, and the number of overlapped reaction cycles.

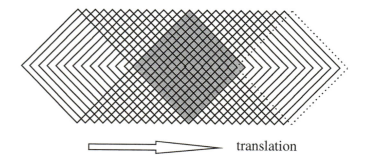

translation

**Fig. 3.12**   The oligonucleotide arrays generated by the successive translation of a diamond-shaped reaction zone. In this hypothetical example, there are a maximum of 13 overlaps leading to a maximum oligonucleotide length of 13 bases. The shaded square contains 169 small squares each bearing a different oligonucleotide sequence of varying lengths.

In practise, it has been found impossible to generate precise diamond-shaped arrays, but oligonucleotide arrays have been successfully synthesized using an overlapping circular mask, leading to irregular overlapped regions. The arrays were easy to fabricate and a semi-automated process was instigated to deliver the reaction solutions.

The original VLSIPS paper described above (Fodor *et al.* 1991) also mentioned the preparation of oligonucleotides on the solid support and highlighted the possibility of making all 65 536 possible octanucleotides ($4^8$) on 50 $\mu$m sites, using a support of 1.6 cm$^2$ total area. Papers published subsequently have realized and surpassed this dream. Workers at Affymetrix, an associated company to Affymax have now generated arrays of 256 different octanucleotides on a chip 1.28 cm $\times$ 1.28 cm (Pease *et al.* 1994). In a similar fashion to the peptide example above the oligonucleotides were synthesized attached to the glass support using 3′-*O*-phosphoramidite-activated deoxynucleotides (see Fig. 3.13). These were protected on the 5′-hydroxyl with a photolabile group (developed specifically for this work), so that the masking technique would permit selective removal of protecting groups in certain regions of the support. The octanucleotides were produced in just four hours following a 16 chemical reaction cycle process.

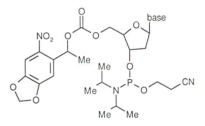

**Fig. 3.13** The protected deoxynucleotides used for oligonucleotide synthesis on glass surfaces.

The library synthesized consisted of all of the 256 oligonucleotide sequences, 3′-CG(A + G + C + T)$^4$CG. When hybridized with the target 5′-GCGGCGGC–fluorescein oligonucleotide, the complementary probe sequence, 3′-CGCCGCCG yielded the most intense fluorescent image (Fig. 3.14), although significant intensity (albeit at least fivefold weaker) was observed for nine mismatches whose probe sequence varied at just one residue.

The authors recognized that although their initial arrays were modest in terms of the numbers of sequences displayed, the technique would soon prove to be a valuable tool for rapid investigation in human genetics and diagnostics. Just two years later, the same company, Affymetrix, in collaboration with the University of California at San Diego published the results of the first clinical application of high density oligonucleotide arrays.

The intention of this study was to examine the DNA sequences of the gene for a key protease in the HIV-1 virus, looking for naturally occurring mutations of the gene that may confer resistance to HIV-1 protease inhibitor drugs. The sequences of protease genes from 167 viral isolates were determined (Kozal *et al.* 1996) using an array just 1.28 cm $\times$ 1.28 cm in size. This small chip contained 12 224 oligonucleotide probe sequences constructed in such a way as to allow the checking of a 382 base pair contiguous sequence in the RNA derived from the HIV-1 protease gene. Each position in the RNA sequence was checked by 16 oligonucleotides on the support, each representing varying lengths (11, 14, 17, or 22 bases) and varying one key base position (A, C, G, or T) in the centre of each probe. Those probes that

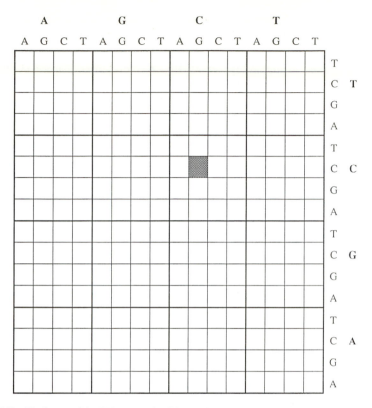

**Fig. 3.14**    The layout of the 256 octanucleotides on the glass support. The site bearing the complementary sequence 3′-CGCGCCCG is highlighted.

were complimentary to the RNA test sequence hybridized most strongly. Examining which probe binds best (as determined by the fluorescence of the fluorescein-labelled RNA) defined the bases in each position of the RNA.

The study revealed that the proteases from these 167 HIV-1 isolates were highly variable, with variation possible in 47 of the amino acids in the 99 residue enzyme sequence. Many of these mutations contribute resistance to the clinically tested protease inhibitors currently being developed for the therapy of HIV infection. It is remarkable to consider that each derivatized region of the glass chip was just 95 $\mu$m × 93 $\mu$m, and contained just $10^8$ molecules of the probe sequence. However, the search is still on for arrays containing even higher numbers of oligonucleotides, and one recent report describes the construction of arrays with around one million sequences per square centimetre (McGall *et al.* 1996).

It was found that the secret of higher density arrays was to avoid the use of photolabile protecting groups, but instead to use semiconductor resists, the methodology used for the manufacture of semiconductors themselves. Glass substrates were derivatized with *N,N-bis*(hydroxyethyl)aminopropyltriethoxy-silane, followed by a 4,4′-dimethoxytrityl (DMT)-protected hydroxy linker and oligonucleotides were prepared on the deprotected hydroxyl group using standard DNA synthesis protocols. The derivatized glass support was covered in a photoresist of an epoxy resin plus a photoacid generator (triphenyl-sulfonium hexafluoroantimonate). The masking procedure allowed pre-defined regions of the support to be exposed to light, and the photogenerated acid polymerized the resin to an insoluble layer. The non-irradiated regions of the support could be deprotected with acid and derivatized with the next nucleotides in the synthesis. This polymeric photoresist process gave very high resolution of structural features on the support using standard oligonucleotide synthesis methodology. The production of 8 $\mu$m features was demonstrated in the paper, and optimization was aiming to produce oligonucleotide arrays with features of below 1 $\mu$m.

Such sensitivity offered by this method of DNA and RNA sequencing offers exceptional promise for future diagnostic clinical uses, and indicates a dramatic future for this most esoteric of combinatorial chemistry methods.

## 3.2   Paper

Any solid phase is suitable for chemistry as long as it offers chemical functionality for building compounds upon, and the material has sufficient stability to the range of chemical conditions likely to be used in synthesis. Cellulose, as in paper or cotton, is a good solid phase for library synthesis as it has a high level of functionality through the numerous available hydroxyl groups. Paper is acid-labile however, rapidly disintegrating when subjected to strong acids as the hydrogen bonding between the oligosaccharide chains is disrupted. Peptide synthesis is possible if the Fmoc protecting group strategy is used, as these protecting groups are removed with strong organic bases and the use of acid can be avoided.

The simultaneous synthesis of different peptide sequences has been achieved on paper disks packed into a multicolumn continuous flow device (Frank and Döring 1988). The peptides were linked to the celluose through *p*-alkoxybenzyl ester linkage which permitted cleavage at the end of the synthesis by a mild treatment with trifluoroacetic acid. The celluose disks are mechanically stable enough to load into and unload from the columns and could be individually labelled by writing on them in pencil. By combining disks into columns, it was possible to simultaneously add the same amino acid to several growing peptides on separate disks, although considerable manual manipulation is required to load and unload the reactor columns.

Having demonstrated that peptides can be synthesized on cellulose disks, a further step in the technology was to recognize that instead of preparing peptides on separate pieces of paper, the compounds could be prepared simultaneously on distinct areas of the same paper sheet. In dispensing small volumes of reagents onto the paper, small circular spots form on the paper, giving rise to the description, 'spot-synthesis' (Frank 1992, 1993). Volumes of reagents used and the positioning of the spots was controlled so that the spots covered much of the paper surface without overlapping. For example, using Whatman 1 'Chr' paper, 0.1 $\mu$l volumes gave rise to 3 mm diameter spots with a total functionality of 0.02 $\mu$mol. Using a thicker paper (Whatman 3MM) up to 50 $\mu$l could be applied to give a 22 mm spot containing 4.4 $\mu$mol of material. The reagents could be applied either manually with volumes down to 0.1 $\mu$l, or by the use of a commercial x,y programmable solution dispenser. As with the cellulose disks, the spots on the paper membrane could be readily and indelibly labelled by pencil.

Combinatorial libraries on paper support have been prepared in an 8 × 12 matrix reflecting the standard layout of the 96-well microtitre plate. The linking to the paper was dictated by the ultimate aims of the library,

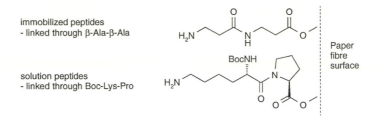

immobilized peptides
- linked through β-Ala-β-Ala

BocNH

solution peptides
- linked through Boc-Lys-Pro

Paper
fibre
surface

**Fig. 3.15**   Two possible linkers used for the synthesis of peptides on paper.

a *β*-Ala–*β*-Ala linker esterified to the paper for peptides to be assayed on the solid phase, or a Boc–Lys–Pro linker if the peptide was to be released by acid catalysed diketopiperazine formation (Fig. 3.15).

With the linkers attached to the paper, peptide synthesis proceeded by the addition of solutions of the appropriate activated amino acids. One useful feature incorporated into the technique was the staining of the sheets with bromophenol blue indicator to show which spots contained unreacted amino groups. The operator could thereby monitor the time course of individual reactions by the colour change from blue to yellow as the free amines were consumed by acylation. Slow reactions could be assisted by further addition of the activated amino acid. Acylation could be terminated by washing the sheet with dilute acetic anhydride in DMF, before commencing the next cycle of synthesis.

The spot-synthesis technique was used in the production of a combinatorial library of overlapping decapeptides derived from CMV26, an immunogenic region of the human cytomegalovirus 36/40K protein. The objective of the study was to determine which parts of this 58 amino acid sequence is recognized by polyclonal antiserum to this protein. Identifying the binding epitopes reveal much about the recognition process between proteins and antibodies.

The library of 49 decapeptides were produced on a *β*-Ala–*β*-Ala linker derivatized paper membrane, and the sheet probed with the rabbit anti-CMV26 serum using an ELISA (enzyme-linked immunosorbant assay). Active sequences were detected by a colorimetric change (Fig. 3.16) and the binding decapeptides were found to contain the epitope sequences, SLSSL and LDNDLMN (Fig. 3.17).

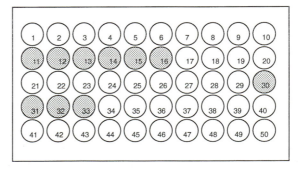

**Fig. 3.16**   The 49 decapeptides were spotted onto a 5 × 10 grid on the paper membrane, and reactive peptides were detected by a colorimetric change.

```
10        20        30        40        50
IEGRGKSRGGGGGGGGSLSSLANAGGLHDDGPGLDNDLMNEPMGLGGLGGGGGGGGKKH
```

```
 1.IEGRGKSRGG    18.SSLANAGGLH    34.DNDLMNEPMG
 2.EGRGKSRGGG    19.SLANAGGLHD    35.NDLMNEPMGL
 3.GRGKSRGGGG    20.LANAGGLHDD    36.DLMNEPMGLG
 4.RGKSRGGGGG    21.ANAGGLHDDG    37.LMNEPMGLGG
 5.GKSRGGGGGG    22.NAGGLHDDGP    38.MNEPMGLGGL
 6.KSRGGGGGGG    23.AGGLHDDGPG    39.NEPMGLGGLG
 7.SRGGGGGGGS    24.GGLHDDGPGL    40.EPMGLGGLGG
 8.RGGGGGGGSL    25.GLHDDGPGLD    41.PMGLGGLGGG
 9.GGGGGGGSLS    26.LHDDGPGLDN    42.MGLGGLGGGG
10.GGGGGGSLSS    27.HDDGPGLDND    43.GLGGLGGGGG
11.GGGGGSLSSL    28.DDGPGLDNDL    44.LGGLGGGGGG
12.GGGGSLSSLA    29.DGPGLDNDLM    45.GGLGGGGGGG
13.GGGSLSSLAN    30.GPGLDNDLMN    46.GLGGGGGGGG
14.GGSLSSLANA    31.PGLDNDLMNE    47.LGGGGGGGGK
15.GSLSSLANAG    32.GLDNDLMNEP    48.GGGGGGGGKK
16.SLSSLANAGG    33.LDNDLMNEPM    49.GGGGGGGKKHH
17.LSSLANAGGL
```

**Fig. 3.17** The analysis of the 49 overlapping decapeptides produced to scan the CMV26 sequence. Peptides that bound to the antiserum are highlighted in bold, leading to identification of the key epitope sequences, SLSSL and LDNDLMN.

The spot-synthesis technique was proven to be an easy way of making peptide combinatorial libraries of sufficient quality for epitope studies. The method is comparable with both the scanning approach on grafted multipins and the much more sophisticated VLSIPS technology. However, its simplicity of approach makes it much more accessible for immediate use. The limitations in the technique lie primarily in the solid phase itself, in that paper is rather limited in the range of chemistry that can be applied without running into handling difficulties, either through the acid-lability or the hydrophilicity of cellulose. Finally, the method does not allow for a convenient release of product peptides for solution assays, although it is feasible that the sheets could be cut up and the individual spots placed into separate tubes for product cleavage.

## 3.3   Laminar polymers

Paper is a good solid phase for a peptide synthesis as the hydrophilic nature of cellulose allows good compatibility with the production of polar peptides. However, there is a limit to the range of chemistry that is compatible with paper. Consequently, there have been attempts to find novel, laminar solid phases that offer compatibility with a wide range of different synthetic conditions. One solution to this problem has been offered by a group at the University of Oxford (Gao and Esnouf 1996*a*). A sheet of Immobilon AV-1 membrane, a chemically inert polyvinylidene fluoride, was derivatized with ethylene diamine to provide a functionalized surface of amine groups. Activated Fmoc–$\beta$-alanine was then spotted onto the membrane at regular spacing to provide spots for the synthesis of peptides. Thus this method is remarkably similar to the paper spots technique of Frank with the key modification in the nature of the membrane solid phase.

The membrane was used to synthesize an immobilized positional scanning combinatorial library of octapeptides prepared by the addition of single Fmoc–amino acids or equimolar mixtures of Fmoc–amino acids and coupling reagents (HOBt and *N,N′*-diisopropylcarbodiimide) in *N*-methyl pyrrolidine. As with the spots-synthesis, the membrane was extensively

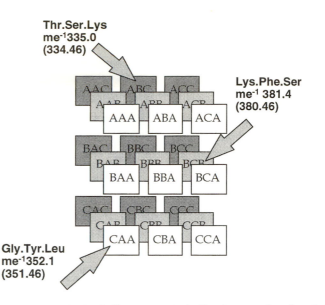

**Fig. 3.19**  Following the synthesis, the library was attached to a large number of small pieces of laminar; each piece containing a unique tripeptide sequence. The presence of each compound was confirmed by identification of the molecular ion by MS (expected value of m/e in parenthesis).

## References

Bray, A.M., Maeji, N.J., and Geysen, H.M. (1990). The simultaneous multiple production of solution phase peptides; assessment of the Geysen method of simultaneous peptide synthesis. *Tetrahedron Lett.*, **31**, 5811–14.

Bray, A.M., Maeji, N.J., Valerio, R.M., Campbell, R.A., and Geysen, H.M. (1991*a*). Direct cleavage of peptides from a solid support into aqueous buffer. Application in simultaneous multiple peptides synthesis. *J. Org. Chem.*, **56**, 6659–66.

Bray, A.M., Maeji, N.J., Jhingran, A.G., and Valerio, R.M. (1991*b*). Gas phase cleavage of peptides from a solid-support with ammonia vapour. Application in simultaneous multiple peptide synthesis. *Tetrahedron Lett.*, **32**, 6163–6.

Bray, A.M., Jhingran, A.G., Valerio, R.M., and Maeji, N.J. (1994). Simultaneous multiple synthesis of peptides amides by the multipin method. Application of vapor-phase ammonolysis. *J. Org Chem.*, **59**, 2197–203.

Bray, A.M., Chiefari, D.S., Valerio, R.M., and Maeji, N.J. (1995). Rapid optimization of organic reactions on solid phase using the multipin approach: synthesis of 4-aminoproline analogues by reductive amination. *Tetrahedron Lett.*, **36**, 5081–4.

Boojamra, C.G., Burow, K.M., Thompson, L.A., and Ellman, J.A. (1997). Solid-phase synthesis of 1,4-benzodiazepine-2,5-diones. Library preparation and demonstration of synthetic generality. *J. Org. Chem.*, **62**, 1240–56.

Bunin, B.A., Plunkett, M.J., and Ellman, J.A. (1994). The combinatorial synthesis and chemical and biological evaluation of a 1,4-benzodiazepine library. *Proc. Natl. Acad. Sci. USA*, **91**, 4708–12.

Fodor, S.P.A., Read, J.L., Pirrung, M.C., Stryer, L., Lu, A.T., and Solas, D. (1991). Light-directed, spatially addressable parallel chemical synthesis. *Science*, **251**, 767–73.

Frank, R. (1992). Spot-Synthesis: An easy technique for the positionally addressable, parallel chemical synthesis on a membrane support. *Tetrahedron*, **48**, 9217–32.

Frank, R. (1993). Strategies and techniques in simultaneous solid phase synthesis based on the segmentation of membrane type supports. *Bioorg. Med. Chem. Lett.*, **3**, 425–30.

Frank, R. and Döring, R. (1988). Simultaneous multiple peptide synthesis under continuous flow conditions on cellulose paper discs as segmental solid supports. *Tetrahedron*, **44**, 6031–40.

Gao, B. and Esnouf, M.P. (1996*a*). Elucidation of the core residues of an epitope using membrane-based combinatorial peptide libraries. *J. Biol. Chem.*, **271**, 24634–8.

Gao, B. and Esnouf, M.P. (1996*b*). Multiple interactive residues of recognition—elucidation of discontinuous epitopes with linear peptides. *J. Immunol.*, **157**, 183–8.

Geysen, H.M. and Mason, T.J. (1993). Screening chemically synthesized peptide libraries for biologically-relevant molecules. *Bioorg. Med. Chem. Lett.*, **3**, 397–404.

Geysen, H.M., Meloen, R.H., and Barteling, S.J. (1984). Use of peptide synthesis to probe viral antigens for epitopes to a resolution of a single amino acid. *Proc. Natl. Acad. Sci. USA*, **81**, 3998–4002.

Geysen, H.M., Barteling, S.J., and Meloen, R.H. (1985). Small peptides induce antibodies with a sequence and structural requirement for binding antigen comparable to antibodies raised against the native protein. *Proc. Natl. Acad. Sci. USA*, **82**, 178–82.

Geysen, H.M., Rodda, S.J., Mason, T.J., Tribbick, G., and Schoofs, P.G. (1987). Strategies for epitope analysis using peptide synthesis. *J. Immunol. Methods*, **102**, 259–74.

Kozal, M.J., Shah, N., Shen, N., Yang, R., Fucini, R., Merigan, T.C., *et al.* (1996). Extensive polymorphism observed in HIV-1 clade B protease gene using high-density oligonucleotide arrays. *Nat. Med.*, **2**, 753–9.

Luo, K., Zhou, P., and Lodish, H.F. (1995). The specificity of the transforming growth factor $\beta$ receptor kinases determined by a spatially addressable peptide library. *Proc. Natl. Acad. Sci. USA*, **92**, 11761–5.

Maskos, U. and Southern, E.M. (1992*a*). Parallel analysis of oligodeoxyribonucleotide (oligonucleotide) interactions. I. Analysis of factors influencing oligonucleotide duplex formation. *Nucleic Acids Res.*, **20**, 1675–8.

Maskos, U. and Southern, E.M. (1992*b*). Oligonucleotide hybridisations on glass supports: a novel linker for oligonucleotide synthesis and hybridisation properties of oligonucleotides synthesized *in situ. Nucleic Acids Res.*, **20**, 1679–84.

McGall, G., Labadie, J., Brock, P., Wallraff, G., Nguyen, T., and Hinsberg, W. (1996). Light-directed synthesis of high-density oligonucleotide arrays using semiconductor photoresists. *Proc. Natl. Acad. Sci. USA*, **93**, 13555–60.

Pease, A.C., Solas, D., Sullivan, E.J., Cronin, M.T., Holmes, C.P., and Fodor, S.P.A. (1994). Light-generated oligonucleotide arrays for rapid DNA sequence analysis. *Proc. Natl. Acad. Sci. USA*, **91**, 5022–6.

Pirrung, M.C. (1997). Spatially addressable combinatorial libraries. *Chem. Rev.*, **97**, 473–88.

Southern, E.M., Case-Green, S.C., Elder, J.K., Johnson, M., Mir, K.U., Wang, L., *et al.* (1994). Arrays of complementary oligonucleotides for analysing the hybridisation behaviour of nucleic acids. *Nucleic Acids Res.*, **22**, 1368–73.

Terrett, N.K., Gardner, M., Gordon, D.W., Kobylecki, R.J., and Steele, J. (1997). Drug discovery by combinatorial chemistry—the development of a novel method for the rapid synthesis of single compounds. *Chem. - Eur. J.*, **3**, 1917–20.

Valerio, R.M., Bray, A.M., Campbell, R.A., DiPasquale, A., Margellis, C., Rodda, S.J., *et al.* (1993). Multipin peptide synthesis at the micromole scale using 2-hydroxyethyl methacrylate grafted polyethylene supports. *Int. J. Pept. Protein Res.*, **42**, 1–9.

Valerio, R.M., Bray, A.M., and Maeji, N.J. (1994). Multiple peptide synthesis on acid-labile handle derivatized polyethylene supports. *Int. J. Pept. Protein Res.*, **44**, 158–65.

Valerio, R.M., Bray, A.M., and Stewart, K.M. (1996*a*). Multipin solid-phase synthesis of acyl 2,3-diaminopropionic acid oligomers. *Int. J. Pept. Protein Res.*, **47**, 414–18.

Valerio, R.M., Bray, A.M., and Patsiouras, H. (1996*b*). Multipin solid-phase synthesis of ethers using modified Mitsunobu chemistry. *Tetrahedron Lett.*, **37**, 3019–22.

Virgilio, A.A. and Ellman, J.A. (1994). Simultaneous solid-phase synthesis of β-turn mimetics incorporating side-chain functionality. *J. Am. Chem. Soc.*, **116**, 11580–1.

Wang, J-X., Bray, A.M., DiPasquale, A.J., Maeji, N.J., and Geysen, H.M. (1993). Application of the multipin peptide synthesis technique for peptide receptor binding studies: substance P as a model system. *Bioorg. Med. Chem. Lett.*, **3**, 447–50.

# 4 Solution phase library synthesis

## 1. Introduction

Much of this book has so far described the synthesis of combinatorial libraries on solid phase. The benefit of preparing libraries on resin beads has been explained as offering advantages in handling, especially where there is a need to separate excess reagents from the reaction products attached to the resin. In most cases a simple filtration effects a rapid purification, and the products are ready for further synthetic transformation. But it should be remembered that using solid phase chemistry brings several disadvantages as well. Clearly the range of chemistry available on solid phase, although rapidly growing, is limited. Furthermore, it is much more difficult to monitor the progress of reactions when the substrate and product are attached to solid phase. Although methods are rapidly being developed to increase the range of solid phase chemistry, and there is an increasing use of IR and NMR techniques to characterize compounds whilst still attached to the solid phase, it is not inappropriate to ask whether all these difficulties could be simply avoided in the first place by using solution phase chemistry to prepare combinatorial libraries.

Indeed, solution phase chemistry has been used for a number of different library synthesis techniques. The difficulty in purifying large numbers of compounds without sophisticated automated processes has in general restricted the use of solution chemistry to short synthetic sequences using reliable chemistry. Additionally, chemists have generally avoided the preparation of mixtures in solution because of the greater potential for the build-up of numerous impurities that would be impossible to remove.

## 2. Parallel solution synthesis

Manual or automated approaches can be used for the parallel preparation of tens to hundreds of analogues of a biologically active substrate. The products are synthesized using reliable coupling and functional group interconversion chemistry and are progressed to screening after removal of solvent and volatile by-products. Parallel and orthodox synthesis is compared in Fig. 4.1.

Orthodox synthesis usually involves a multistep sequence, e.g. from **A** through to the final product **D**, which is purified and fully characterized before screening. The next analogue is then designed, guided by the biological activity of the previous compound, prepared, and then screened. This process is repeated to optimize both activity and selectivity. In contrast parallel analogue synthesis involves reaction of a substrate **S** with multiple reactants, $\mathbf{R}^1$, $\mathbf{R}^2$, $\mathbf{R}^3$.....$\mathbf{R}^n$, to produce a compound library of **n** individual products $\mathbf{SR}^1$, $\mathbf{SR}^2$, $\mathbf{SR}^3$.....$\mathbf{SR}^n$. The library is screened, usually without purification,

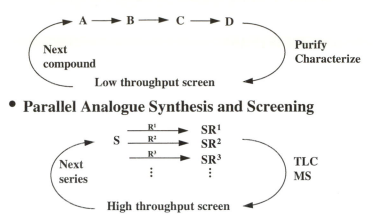

**Fig. 4.1**   The comparison of orthodox and parallel analogue synthesis.

and with only minimal characterization of the individual compounds, using a rapid throughput screening technique. If any active compounds are identified, they are resynthesized on a larger scale for purification, characterization, and screening by traditional methods. If biological activity is confirmed, the newly discovered leads and SAR are used to design new substrate templates. Further libraries are then prepared, again using the parallel analogue synthesis methodology, now focused on the new substrate. This process is thus a rapid and iterative fine-tuning similar to conventional lead optimization, only much faster and involving many more analogues.

Typically, parallel synthesis involves one- or two-step reactions using reliable solution chemistry such as reductive aminations, acylations, and Suzuki couplings performed on a relatively small scale ($\sim$10 $\mu$moles). Initially, trial reactions are carried out to optimize the reaction conditions, and this also dictates the vessels to be used (e.g. Reacti-vials$^{TM}$, 96-well plates, microwave vials). The reactant library (a selected set of reactants with appropriate functional groups, e.g. acid chlorides) necessary for the desired analogues is assembled, usually from commercial sources. The reactant set can either be targeted, using available SAR information, or non-targeted but selected to encompass a diverse range of physical and chemical properties, such as molecular size ($M_r$), lipophilicity (Log D), and acidity ($pK_a$). Once selected, the reactants and substrate are dissolved in a suitable, usually volatile solvent (e.g. dichloromethane, acetonitrile, THF) to give solutions of known concentration. The required volume of the substrate solution and each of the reactant solutions is dispensed into the reaction vessels, along with the required quantity of any necessary reagent solutions. These reaction mixtures are subjected to appropriate reaction conditions, such as heating or sonication, following which the crude products can be analysed by thin-layer chromatography (TLC) and MS to give an indication of the average extent of reaction and percentage of products formed. Solid phase reagents such as polymer-bound coupling agents are used where possible since they can enhance product purity, due to the ease of separation from products using a simple filtration step.

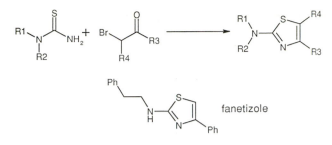

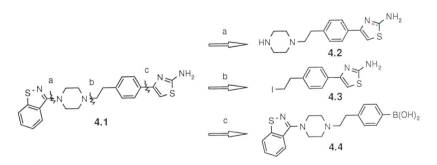

**Fig. 4.2**  The Hantzsch synthesis of aminothiazoles and the structure of fanetizole.

A typical solution phase library is described by workers from Glaxo–Wellcome (Bailey *et al.* 1996). A library of 20 2-aminothiazoles as individual compounds was prepared using the reliable and traditional Hantzsch synthesis, in a 4 × 5 grid of one dram glass vials (Fig. 4.2). The compounds were prepared by the mixing of five thioureas separately with four α-bromoketones, using a liquid dispensing robot to transfer the solutions. The reactions were heated at 70 °C for five hours before quenching with dimethylamine. Following completion of the synthesis, the DMF solvent was removed by blowing a stream of nitrogen gas into each vial at ambient temperature for about 24 hours.

Analysis of the products by HPLC and LC–MS demonstrated that in each case the principle component had a molecular ion corresponding to the expected thiazole. High resolution MS and proton NMR confirmed the identity of the products. The value of the technique on drug discovery was suggested by the synthesis within this set of 20 compounds, of the known anti-inflammatory compound, fanetizole.

Similar parallel synthesis of single compounds has been successfully used in the preparation of analogues of antiviral compounds and neurokinin-2 (NK$_2$) antagonists (Selway and Terrett 1996). Parallel synthesis techniques were used to explore SAR around the antiherpes aminothiazole derivative, **4.1** (Fig. 4.3), in both the herpes simplex virus (HSV-1) helicase adenosine triphosphatase (ATPase) assay, and the antiviral plaque reduction assay (IC$_{50}$ = 3 $\mu$M versus HSV-1 helicase ATPase and 11 $\mu$M antiviral activity in a HSV-1 plaque reduction assay).

Retrosynthetic analysis of **4.1** using the disconnections shown in Fig. 4.3, led to three intermediates, **4.2**, **4.3**, and **4.4**, all of which were candidate substrates for parallel synthesis.

**Fig. 4.3**  Three possible disconnections followed up through parallel synthesis.

Disconnection **a** gave the piperazine derivative **4.2** which reacted readily with electrophiles to give 400 novel analogues. Disconnection **b** led to an iodide **4.3**, which underwent nucleophilic substitution to give 360 analogues, and disconnection **c** suggested the boronic acid intermediate **4.4**, which under Suzuki coupling conditions reacted with heteroaryl halides to form 60 new biaryl derivatives. Previous SAR suggested that the aminothiazole was important for activity, so the first two disconnections retained this heterocycle although the third disconnection allowed replacements to be sought.

The highlight from this study was the piperazine (**4.5**) which, with an $IC_{50}$ value of 0.6 $\mu$M versus HSV-1 helicase ATPase, was fivefold more active than the parent compound.

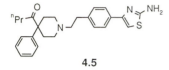

**4.5**

The second example of the use of parallel synthesis for rapid SAR evaluation in this study centred on a known $NK_2$ antagonist, **4.6** (Lawrence *et al.* 1992). This structure presented a number of bonds that suggested obvious reliable one-step reactions (Fig. 4.4).

Parallel synthesis using these substrates gave four compound libraries containing a total of 1180 individual compounds screened against rat duodenal $NK_2$ receptors. Although no compounds exceeding the potency of the original lead were discovered, extensive valuable SAR was derived from these libraries. All of the compounds were prepared using parallel synthesis methodology in approximately six weeks, and screening was carried out in a few days, demonstrating that parallel synthesis can considerably enhance the rate at which medicinal chemists can prepare analogues of lead compounds to determine SAR information.

The parallel solution synthesis of single compounds is a valuable addition to the tools available to the medicinal chemist as it allows the rapid synthesis of many compounds for biological evaluation. Single compounds are especially valuable for lead optimization, as there is a high probability that there will be many active components within the library. In these circumstances a library prepared in mixtures is inappropriate as too many of the mixtures would display activity, and deconvolution would be a lengthy procedure. Testing single compounds thus permits the rapid identification of both active compounds and structure–activity relationship information.

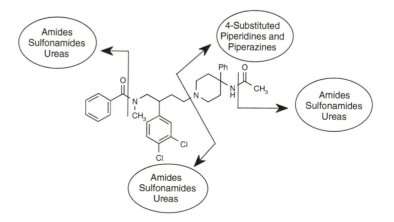

**Fig. 4.4**   Four disconnections for the parallel analogue synthesis of a known $NK_2$ antagonist.

Many research organizations are now using the parallel synthesis of single compounds for their drug discovery programmes, and this technique will continue to grow in importance with the advent of superior automated synthesis and purification procedures.

## 3.   Indexed combinatorial libraries

Libraries of compounds in mixtures have been prepared by using solution techniques. In particular the technique of indexed libraries has been used to prepare such libraries and quickly identify the most biologically active compounds from within the library. This technique, totally analogous to the process of orthoganol combinatorial libraries prepared on solid phase (see Chapter 2, Section 7) was first described by a group from Glaxo (Smith *et al.* 1994) who synthesized a total of 1600 amides and esters through the single-step reactions of 40 acid chlorides with 40 amine and alcohol nucleophiles. The library, presented for screening in just 80 mixtures, contained every compound twice, and activity in any particular combination of mixtures instantly indicated the structure of the most active component.

The library was constructed by taking each of the acid chlorides (A) and reacting it with a stoichiometric amount of an equimolar nucleophilic mixture ($N_{1-40}$) in dichloromethane solution. Similarly, each of the nucleophiles (N) was reacted with a mixture of all the acid chlorides ($A_{1-40}$) (Fig. 4.5). After incubating the reaction mixtures at ambient temperature for 48 hours, methanol was added to quench any remaining acid chlorides and the solvent evaporated in air. Although there was no attempt made to monitor the progress of the reactions in mixtures, the synthesis of representative examples as individual compounds was validated in advance, and some of the mixtures were analysed by both GC and GC–MS.

Thus the library had been prepared twice, and a mixture from the first set of samples active in a biological assay would define the acid chloride component of the active compound. Activity from the second set of mixtures would define the structure of the nucleophile. The library was screened through a large number of different biological assays at between 1–10 $\mu$M (concentration per component), and compounds with weak activity against the neurokinin-3 (NK$_3$) receptor (**4.6**, IC$_{50}$ = 60 $\mu$M) and matrix metalloproteinase-1 (**4.7**, IC$_{50}$ = 55 $\mu$M) were discovered.

The authors found that common to all mixtures synthesized through combinatorial methods, the activity of mixtures in a biological assay tends to overestimate the potency of the final compound. This is a common

| Set 1 (40 samples with A fixed) | Set 2 (40 samples with N fixed) |
|---|---|
| $A_1 + N_{1-40}$ → sample 1 | $N_1 + A_{1-40}$ → sample 1 |
| $A_2 + N_{1-40}$ → sample 2 | $N_2 + A_{1-40}$ → sample 2 |
| \| | \| |
| $A_{40} + N_{1-40}$ → sample 40 | $N_{40} + A_{1-40}$ → sample 40 |

**Fig. 4.5**   The synthesis of the two sets of mixtures in solution that together constituted an indexed library.

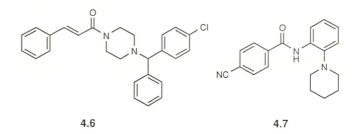

|  |  |
|---|---|
| **4.6** | **4.7** |

consequence of screening a mixture of structurally related compounds. As many of the components may have some weak biological activity the assay result indicates the additive effect of all of these active compounds, rather than the activity due to the most potent component alone.

Another indexed solution library has been described (Pirrung and Chen 1995). In this example, 54 carbamates were prepared from a set of nine alcohols and six isocyanates, and as with the Glaxo example above, the library was constructed twice; one set was constructed from each alcohol reacting with a stoichiometric equimolar mixture of isocyanates, and the second from each isocyanate reacting with an stoichiometric equimolar mixture of alcohols (Fig. 4.6).

The products were screened against electric eel acetylcholinesterase, the activities of the mixtures being used as 'indices' to the rows and columns of a two-dimensional matrix representing the activities of the individual carbamate products. In Fig. 4.7 the activities (expressed as the reciprocal of the enzyme $IC_{50}$) of the 15 mixtures are plotted as columns on the two-dimensional matrix. The intersections of the rows containing the tallest columns (i.e. mixtures 6 and A) should represent the most potent compounds in the library. In this library the single compounds in these two rows were synthesized as individual compounds, and the assay results confirmed that indeed compound 6A (**4.8**) was the most potent of those synthesized.

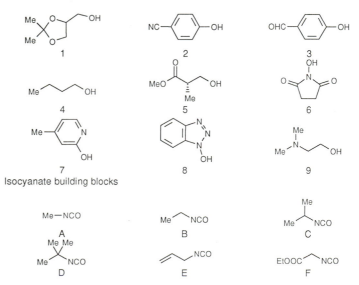

**Fig. 4.6**  The sets of alcohols and isocyanates used in the synthesis of an acetylcholinesterase inhibitor combinatorial library.

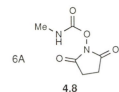

**4.8**

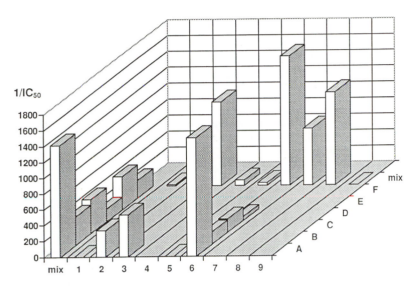

**Fig. 4.7** The activity of the individual mixtures and selected individual compounds from an indexed combinatorial library screened against acetylcholinesterase.

This group has used the same solution indexed library approach to prepare a library of 72 tetrahydroacridines for screening against the same enzyme, electric eel acetylcholinesterase (Pirrung *et al.* 1995). A mixture of 12 cyclohexanones were reacted individually with six *o*-cyanoanilines under acidic conditions to generate one dimension of the indexed library, and the second dimension was generated by reacting the mixture of six anilines with the individual cyclohexanones (Fig. 4.8). The compound common to the most potent mixture in each of the two dimensions was identified as the 7-nitro derivative (**4.9**), which with a $K_i$ value of 10 nM was some tenfold more potent than the parent inhibitor, tacrine (**4.10**).

The indexed library approach, although successful for small libraries may give rise to problems with larger mixtures. First, all of the complications of testing mixtures can occur, with the risk of possible synergistic or antagonistic effects within the biological assay. Assuming that such effects don't occur, the two-dimensional approach to library synthesis still suffers from the limitation

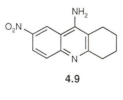

**4.9**

**4.10**

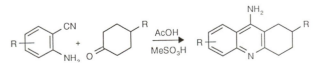

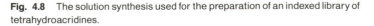

**Fig. 4.8** The solution synthesis used for the preparation of an indexed library of tetrahydroacridines.

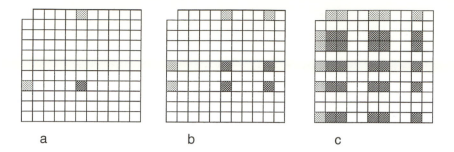

**Fig. 4.9**  A hypothetical library of 100 dimeric compounds prepared from two sets of ten monomers. The top row and the left-hand column represent to the two sets of mixtures that are screened, and the matrix between represent the 100 compounds in the library. Indexed libraries permit the discovery of active compounds by identifying the intersection of active mixtures. (a) If only one mixture in each set is active, the single active compound can be unambiguously determined. (b,c) If more than one mixture in each set reveals interesting biological activity, the identity of the active constituents is more difficult to ascertain, and requires the resynthesis of several potentially active compounds.

that in the absence of redundancy in the library, there is consequently no latitude for error. It is essential that both mixtures containing an active compound appear active in the assay, otherwise it is not possible to find the potent compounds in the library. If the biological assay indicates that one mixture in each set is active, there is no ambiguity, and the compound common to each of these sets is likely to be the most potent (see Fig. 4.9a). However, if there are two mixtures in each set that stand out as potent, there is no way to tell which of the four possible compounds at the intersection points is active, and all four would need to be resynthesized and retested to be able to identify the most potent (Fig. 4.9b).

Furthermore, this problem increases arithmetically. If there are five active mixtures in each set, as many as 25 individual compounds would need to be synthesized to find the best compounds of the library (Fig. 4.9c).

This problem is not only common to indexed libraries prepared in solution. Positional scanning peptide libraries (see Chapter 2, Section 6) are also a form of indexed library although this description is usually reserved for compound libraries that consist of more than two components. It will be recalled that in this technique the same library compounds are prepared several times, and in each library a different residue in a peptide sequence is held constant. For example, a hexapeptide library of 34 million ($18^6$) compounds was produced in six sets of mixtures, each mixture containing 1 889 568 ($18^5$) peptides (Dooley and Houghten 1993). The most potent $\mu$-opioid receptor ligands in this library were found by the resynthesis and screening of 24 individual peptides. Intriguingly the hexapeptide based on the most preferred single residue in each position was only weakly active.

This example was a library prepared on solid phase, an obvious choice given the length of the synthetic routes to hexapeptides. The two solution indexed libraries cited above work well because the chemistry consists of only one synthetic step. If the compounds were larger, consisting of more monomeric units, the synthesis would more likely have been performed on solid phase and the process would instead be described as another example of positional scanning.

# 4. Template-based libraries

The examples above involve the solution synthesis of compounds from two precursor molecules. An approach to the synthesis of a solution-based library where four building blocks are added to a core template has been described (Carell *et al.* 1994*a*). The core molecules employed were the essentially planar 9,9-dimethylxanthene tetracarbonyl chloride (**4.11**) and the tetrahedrally substituted cubane tetracarbonyl chloride (**4.12**).

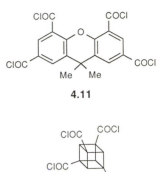

**4.11**

**4.12**

Each of these core molecules was reacted with four equivalents of an amine mixture. An excess of amines was used to drive the reaction to completion, but to prevent unequal quantities of products, the monomers (mainly amino acid derivatives and aminoheterocycles) were chosen to have approximately equivalent reactivity. Washing the reaction solution with citric acid and sodium bicarbonate solutions to remove unreacted starting materials gave the products as powders.

The xanthene core was reacted with four, seven, 12, or 21 amines to give 136, 1225, 10 440, or 97 461 theoretical products respectively. The cubane template, because of its higher order symmetry was expected, using the same amine mixture sizes, to yield 36, 245, 1860, and 16 611 products respectively. HPLC traces of the mixtures demonstrated the expected increasing complexity as the mixture size increased although there was no expectation of a quantification of the pool size. Small xanthene mixtures (171 components) have been characterized by capillary electrophoresis in tandem with mass spectrometry (Dunayevskiy *et al.* 1996) (see Chapter 8, Section 4.1).

In a separate experiment the two core molecules were each reacted with 19 amino acids to yield 65 341 products from the xanthene and 11 191 products from the cubane, and these mixtures were assayed in the search for an inhibitor of the trypsin catalysed hydrolysis of *N*-$\alpha$-benzoyl-D,L-arginine-*p*-nitroanilide (BAPA) (Carell *et al.* 1994*b*). Only the library generated from the xanthene caused a significant reduction in trypsin activity. To identify the amine residues that conferred greatest trypsin affinity in this mixture, three further screening steps were carried out. The amino acids (omitting cysteine as it is known to cause artefactual results in some enzyme assays) were split into six groups of three, and the mixture regenerated using only five of these groups (Table 4.1). Testing these six mixtures indicated that when certain groups of amino acid were omitted from the mixture, the ability of the library to inhibit trypsin was reduced. Specifically, the amino acids in groups G1, G2, and G3 conferred the greatest trypsin affinity to the xanthene-based library and were used exclusively for further library analysis.

Further rounds of synthesis omitting amino acids in turn finally indicated that highest trypsin inhibitory activity resulted from the inclusion of Lys, Ile,

**Table 4.1** The six groupings of three amino acids used in the synthesis of a library of trypsin inhibitors

| G1 | G2 | G3 | G4 | G5 | G6 |
|----|----|----|----|----|----|
| Gly | Leu | Arg | Ser | Phe | Glu |
| Ala | Ile | Lys | Thr | Tyr | Asp |
| Val | Pro | His | Met | Trp | Asn |

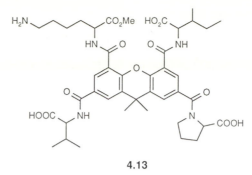

4.13

Pro, and Val residues. The 12 possible structural isomers of the xanthene substituted by these four amino acids were prepared and a 9 $\mu$M trypsin inhibitor (**4.13**) was finally identified.

The process by which this compound was discovered does not rule out the possibility that there are other potent compounds within the original library mixtures. A change in screening strategy as trivial as changing the initial groupings of the 18 amino acid building blocks may have led to an entirely different final active compound. A computer program has been written to simulate activities for all of the sublibraries (Carell *et al.* 1995), calculating the binding contribution made by each building block in each of the positions around the xanthene template. The conclusion of this study was that it is likely that the compound discovered is one of the most active trypsin inhibitors in the initial library, although it is theoretically possible with this screening strategy to overlook another active inhibitor with similar or even slightly higher enzyme affinity.

## 5.  Liquid phase combinatorial synthesis

Solution library synthesis offers the standard advantages of orthodox chemistry in that there is no necessity to develop a new range of chemistry to prepare libraries. However, as previously discussed, the uncertainties of solution chemistry demands some form of purification after each step in the synthesis; a difficult process if many compounds are being synthesized in parallel and impossible if mixtures are being created. Thus in the examples above, the chemistry is generally just one-step with no attempt at purification.

However, there have been some successes at marrying together the advantages of orthodox solution organic chemistry with the simple purification by filtration offered by solid phase synthesis. Liquid phase combinatorial synthesis is a method of preparing libraries chemically bound to a substance that can be in solution or precipitated as a solid on command (Han *et al.* 1995). The key feature is the use of a soluble, linear homopolymer, polyethylene glycol monomethyl ether (MeO–PEG) which serves as a protecting group on the synthesized compound. In many solvents the polymer has a solubilizing effect keeping the library substrate in solution during reaction. The mix and split protocol can be carried out whilst the library is in solution and homogeneous, allowing a more accurate division of the library. However, in diethyl ether, MeO–PEG has a strong propensity to crystallize allowing the library products to be purified by washing away unwanted excess coupling reagents in a fashion identical to the washing of resin beads.

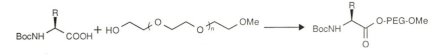

**Fig. 4.10** The attachment of Boc-protected amino acids to the polyethylene glycol solution polymer.

This method has been used in the synthesis of a library of peptides screened for binding to the monoclonal antibody raised against $\beta$-endorphin. A peptide library of 1024 pentapeptides constructed from the amino acids, Tyr, Gly, Phe, and Leu was produced with MeO–PEG attached through an ester linkage to the carboxy terminus of the growing peptide chain (Fig. 4.10). The library mixtures were screened whilst still attached to the MeO–PEG polymer, and through the process of recursive deconvolution, the following several potent binding sequences were determined.

Structure:  Tyr–Gly–Gly–Phe–Leu–PEG–OMe   $IC_{50} = 34$ nM
            Tyr–Gly–Gly–Phe–Phe–PEG–OMe               49 nM
            Tyr–Gly–Gly–Phe–Tyr–PEG–OMe               91 nM

## 6.  Dendrimer-supported combinatorial chemistry

Related to liquid phase combinatorial synthesis is a recent development in which dendrimers support combinatorial chemistry (Kim *et al.* 1996). Dendrimers are branching oligomers characterized by discrete and controllable molecular architectures. The principle of building compound libraries on the dendrimers provides several advantages. Solution phase techniques may be used, although purification may be effected by size-based separation techniques such as size exclusion chromatography (Fig. 4.11). Furthermore the intermediates may be characterized by a variety of routinely available analytical methods such as NMR, IR, UV and MS. Finally, because of the nature of dendrimer structure, the molecules provide extremely high loading.

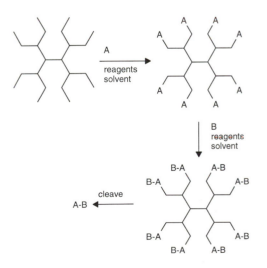

**Fig. 4.11** The nature of dendrimer structure, and a schematic representation of building the compound AB on a soluble dendrimer support.

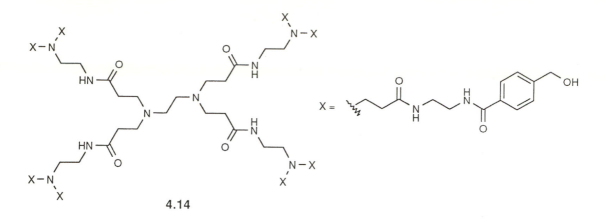

**4.14**

For example, only 7 mg of the dendrimer (**4.14**) are required to obtain the same amount of product that would be synthesized on 100 mg of a resin with loading of 0.23 mmol/g. The dendrimer structure (**4.14**) has been used for the synthesis of a library of indoles constructed using the Fischer synthesis giving individual compounds in high purity and yield. Like solid phase synthesis, a chemical cleavage step at the end of the synthesis provides the product compounds free of the supporting dendrimer.

Dendrimer combinatorial chemistry offers an alternative strategy for solution phase library synthesis which is still at a very early stage of development. The ability to tailor-make dendrimers with specific properties of solubility, chemical stability, and loading capacity allows ready optimization of the library synthesis process.

## 7.   Solution phase libraries with liquid–liquid purification

Unlike synthesis on solid phase, it is not possible to purify library compounds produced in solution by using a simple filtration. Thus, apart from short, high yielding and reliable synthetic routes, solution phase library synthesis has been passed over in favour of solid phase methods. However, an easy way to separate products from intermediates can be achieved by exploiting some physical property of the compounds being prepared. For example, the preparation of compounds based on imino diacetic acids allows purification and product isolation after each step by simple liquid–liquid extraction (Cheng *et al.* 1996*a*). A 27 component library was constructed as a 3 × 3 × 3 matrix of individual compounds following the synthesis in Fig. 4.12 and the

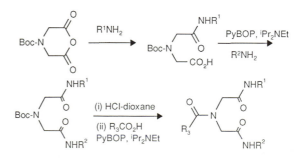

**Fig. 4.12**   The solution synthesis of iminodiacetic acid libraries.

**Table 4.2** The monomers used in the synthesis of a iminodiacetic acid library

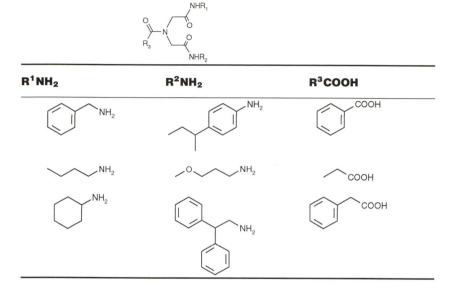

| R¹NH₂ | R²NH₂ | R³COOH |
|---|---|---|

monomers listed in Table 4.2. After each reaction the product could be isolated by appropriate use of acid and/or base washes, leading finally to products in 90–100% purity and 9–84% overall chemical yield.

The same approach has been used to synthesize a library based on a bicyclic pyrrole system (Boger *et al.* 1996). This rigid system (Fig. 4.13) in its fully extended form was proposed as a Gly–X dipeptide mimic. The starting anhydride (**4.15**) was reacted with nucleophiles (e.g. amines, alcohols, or thiols) to generate a new bond and simultaneously to reveal a carboxylic acid as a second functionalization site. The carboxylic acid also permitted the separation of the product from other materials through a liquid–liquid or solid–liquid extraction. The acid could be derivatized, followed by deprotection and acylation of the amine. The basic amine also provided a convenient handle for a second separation step. This template has been used for the solution synthesis of a 27 component (3 × 3 × 3) library using monomers similar to those listed in Table 4.2 above. The compounds were obtained as fully characterized individual samples in equal to or greater than 90% purity.

In these two examples, where a simple isolation procedure has replaced the filtration used in solid phase synthesis, the full repertoire of solution chemistry is available for library synthesis, with the proviso that a suitable functional group is exposed at appropriate points through the synthesis. This approach is

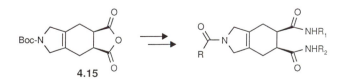

**4.15**

**Fig. 4.13** Derivatization of the cyclic amino anhydride was used to give a 27 component library of triamides.

not limited by scale or the need to provide a linking functional group for attachment to the solid support.

## 8.   Fluorous solution chemistry

As phase separation is the key to exploiting reliable solution phase chemistry for library synthesis, where might this technology proceed in the future? One possible direction is the use of fluorous reagents and labels for library synthesis (Curran 1996). The low polarizability of fluorine leads to very weak van der Waals forces between perfluoro organic compounds. Consequently they are often immiscible with orthodox organic solvents although they will solubilize other perfluoro organic compounds. Thus it should be possible to develop a number of specialist perfluoro reagents for use in solution library synthesis, and these may be readily removed at the end of the synthetic step by a liquid–liquid extraction between an organic and a fluorous solvent. This has been demonstrated for a radical reduction of bromoadamantane using ((2-perfluoro-hexyl)ethyl)tin hydride in trifluoromethylbenzene as solvent (Curran and Hadida 1996). The reaction was worked-up by evaporation of the solvent and partitioning of the product mixture between perfluoromethylcyclohexane and methylene chloride. Evaporation of the organic layer gave adamantane in 90% yield whilst the perfluoro solvent layer contained the tin by-products. Modification of the reaction to give a catalytic reaction by the inclusion of sodium cyanoborohydride, AIBN, and *tert*-butanol required a three-phase liquid extraction work-up. Water (top layer) removed the inorganic products, methylene chloride (middle layer) extracted the product, whilst the bottom layer (perfluoromethylcyclohexane) took the tin by-products.

In another illustration of this approach, Stille couplings have been successfully carried out using a modified fluorous aryltin reactant (Curran and Hoshino 1996). After reaction with a triflate or halide under palladium catalysis, the desired Stille coupling product was extracted into dichloromethane, whilst the fluorous tin by-products were extracted into FC-72 (a fluorocarbon solvent mainly consisting of $C_6F_{14}$ isomers) (Fig. 4.14). In this fashion, a combinatorial library was prepared by reacting each of a number of aryl tin reagents with mixtures of five halides or triflates.

Fluorous reagents may be removed by extraction into fluorous solvents, but the use of a fluorous phase label allows the use of conventional reagents for library synthesis, and fluorous products can be selectively extracted away from non-fluorous regents and unreacted starting materials (Studer and Curran, 1997; Studer *et al.* 1997). Such a label is the silyl bromide (**4.16**) which has fluorinated chains large enough to force labelled compounds into the fluorous phase. Labelling of allyl alcohol using a reaction where the alcohol was present in excess and purification by extraction preceded nitrile *N*-oxide

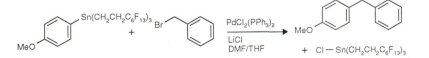

**Fig. 4.14**   An example of a Stille coupling that proceeded with a fluorous aryltin reagent in 98% yield.

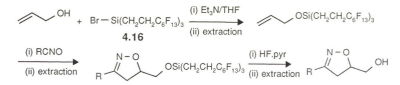

**Fig. 4.15** The use of a fluorous reagent to label the starting material and permit selective extraction away from excess starting materials and reagents.

cycloaddition. The isoxazoline product of this reaction is labelled and could be subsequently extracted into a fluorous solvent away from unreacted starting materials (Fig. 4.15). Fluoride catalysed cleavage of the fluorous label and a further phase separation into three phases provides the pure isoxazoline product through selective extraction into the organic phase.

An alternative procedure is to use a 'phase switch' to purify the products of a solution reaction. The alcohol product of a Grignard reaction has been reacted with a fluorous reagent to switch solvent affinity from organic to fluorous and permit a subsequent three-phase extraction. After this separation, the fluorous label can be cleaved and the pure product extracted selectively into an organic solvent phase. One concern that is being addressed is the size of fluorous label required to render an organic molecule selectively soluble in fluorous solvents. The label molecule, **4.16**, will solubilize organic molecules under 200 but does not have sufficient fluorine content to bring molecules in the molecular weight range 300–500 into fluorous solution.

Overall, the simplicity of this fluorous chemistry holds out much promise for an entire fluorous vocabulary, with specialist perfluoro reagents or fluorous labels designed for high yielding and readily purifiable solution combinatorial chemistry. Excess reagents can be used to push reactions to completion and a simple two-phase liquid–liquid, or three-phase liquid–liquid–liquid extraction will suffice to purify the products. The synthesis will have been designed such that the desired product has a different phase affinity from all other components in the reaction sequence. Such an approach to solution combinatorial chemistry whilst it requires significant investment of resource up-front to design the syntheses and devise the fluorous reaction technologies, lends itself to rapid throughput and, ideally, automation of synthesis and work-up. Solution library synthesis will benefit from further developments of this innovative approach.

## 9.  Solution chemistry with solid phase reagents

There have been recent advances in the use of solid phase reagents to purify compounds made in solution. More than one equivalent of reagent can be used in solution to push a reaction to completion and the excess reactive reagents may be removed by solid phase scavengers (Kaldor *et al.* 1996*a*). Thus as outlined in Fig. 4.16, reacting substrate B with an excess of reagent A will give A–B plus the excess of A. If A is an electrophilic component, a nucleophilic solid phase component such as aminomethyl resin will react with the reagent A and can be removed by filtration.

This idea is not totally new. Reactive impurities have been removed from natural product mixtures using solid phase covalent scavengers (Cheminat

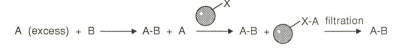

**Fig. 4.16**  The use of a covalent scavenger for the removal of excess reagent A after the solution reaction has been pushed to completion.

*et al.* 1980), but the application of this approach to library synthesis is a useful modern day adaptation. In a variety of examples, the Lilly group demonstrate that nucleophilic covalent scavengers will remove excesses of isocyanates, acyl chlorides, and sulfonyl chlorides, whilst solid phase-bound isocyanates, aldehydes, or acyl chlorides will remove amines. This methodology has been put to good use in the preparation of a solution library of 4000 ureas used in the discovery of novel leads with antirhinoviral activity (Kaldor *et al.* 1996*b*). The urea library was prepared in solution from 1.25 equivalents of single isocyanates reacted with equimolar mixtures of ten amines. Aminomethyl resin was employed to remove the excess isocyanate, and the product mixtures of ten ureas each were screened for their activity against human rhinovirus-14 (HRV-14) in cell culture. Deconvolution of the most active mixtures led to the identification of two antiviral agents active at 1 $\mu$gml$^{-1}$ although with low cytotoxicity (Fig. 4.17).

In a subtle modification of this methodology, ion exchange resins have been used as scavengers to remove excess reagents by salt formation rather than through covalent bond formation (Gayo and Suto 1997). A range of basic ion exchange resins were examined for their ability to remove carboxylic acid by-products from the solution phase reaction of amines with excess acid chlorides. It was discovered that Amberlite® IRA-68, a weakly basic resin provided products of the highest purity (Fig. 4.18). In an analogous fashion, excess amines used in the production of ureas from isocyanates could be removed by the use of acidic ion exchange resins such as Amberlite® IR-120 or Amberlyst 15.

Amberlite® resin has also been used in the synthesis of ethers, although it is difficult in this particular library example to define the synthesis category as solution or solid phase (Parlow 1996). The quaternary ammonium chloride resin was packed into a column and thoroughly basified with sodium

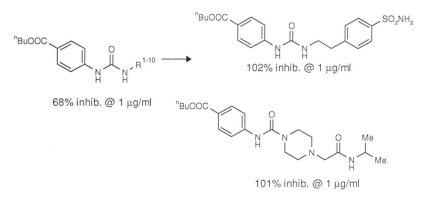

**Fig. 4.17**  The discovery of two antiviral compounds through a solution phase library synthesis using solid phase scavengers to remove excess reagents.

| Resin | Yield (HPLC area) | Purity |
|---|---|---|
| Amberlite® IRA-68 (weakly basic) | 30 | >99% |
| Amberlite® IRA-904 (strongly basic) | 16 | 99% |
| Dowex® 1X2-100 (strongly basic) | 30 | 95% |
| Dowex® 66 (weakly basic) | 16 | 94% |
| AmberJet™ 4200 (strongly basic) | 15 | 92% |
| No resin | 17 | 60% |

**Fig. 4.18**  Results of varying the ion exchange resin added to scavenge excess carboxylic acid. The reactions were all performed in ethyl acetate adding the resin after the reaction was completed. Yields (product areas) and purities were determined by HPLC.

hydroxide before reacting with a mixture of phenols or hydroxyheterocycles. These generated a mixture of polymer-bound quaternary ammonium alkoxides which were reacted with a known amount of *n*-butyl bromide in THF at 65 °C for six hours (Fig. 4.19). Filtration of the resin, washing exhaustively with THF and evaporation gave a mixture of the desired ethers in 95% yield. The use of an anion exchange resin gives the advantages of solid phase reagents but the simplicity of solution chemistry, and for single-step chemistry would be hard to beat for its effectiveness.

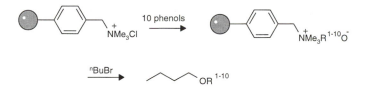

**Fig. 4.19**  The synthesis of a library of ether products on an ion exchange resin.

## 10.  Solution chemistry with resin capture

The section above describes how functionalized solid phases can be used to purify the products of solution phase chemistry. The use of polymeric reagents for phase separation can be taken one step further. Instead of providing functionality for the removal of unwanted by-products, the resin can 'capture' the desired product in the reaction to separate it from the by-products, and even to allow further chemistry to be performed whilst attached to the support.

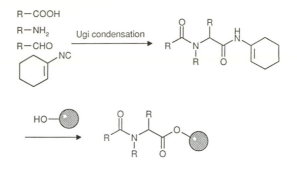

**Fig. 4.20**   The Ugi four-component condensation followed by resin capture of the product onto Wang resin.

An example of this innovative modification of solution phase library chemistry has been demonstrated by Keating and Armstrong (1996). The Ugi four component condensation reaction has been especially successful for combinatorial library synthesis as four components can be condensed simultaneously to give diverse products. The reaction of a carboxylic acid, a primary amine, an aldehyde, and an isocyanide gives versatile amino acid derivatives as products. Armstrong's group have overcome the lack of commercially available isocyanides by employing cyclohexene isocyanide as a 'universal' isocyanide that can be converted to a range of post-Ugi derivatives. In one modification, the Ugi product prepared in solution was captured on Wang resin and was thus neatly separated from the unreacted starting materials and by-products (Fig. 4.20).

Resin capture allows the use of a solution reaction in the first step of a synthetic sequence, and subsequent phase separation by capture onto the solid phase. Further solid phase transformations are now possible, or alternatively the product can be cleaved directly from the resin.

Armstrong has also used resin capture for the synthesis of tetrasubstituted ethylenes related to tamoxifen (Brown and Armstrong 1996). *Bis*(boryl)alkenes were monoarylated with 1.5 equivalents of organohalide under Suzuki conditions in solution to generate both mono- and dialkylated products. Capture with Rink resin-bound aryl iodide in a second Suzuki reaction proceeded without the addition of any further palladium catalyst (Fig. 4.21). Only the monocoupled product is captured onto the resin, as any *bis*-coupled product would no longer have a boronate capable of reacting with the resin-bound aryl iodide. The products were isolated as a mixture of regioisomers in > 95% yield.

## 11.   Solution chemistry using polymeric reagents

One last variation on solution phase chemistry is the use of polymeric solid phase reagents for multistep synthesis. Because reagents are 'locked up' inside the resin bead, the reagents on one bead will not contact, and thus cannot react with, the reagents on another bead. Thus, it is possible to mix acidic and basic resin in solvent without neutralization being able to take place! This observation leads to the possibility that with a suitable choice of polymer reagents and a suitable substrate, it may be possible to carry out a

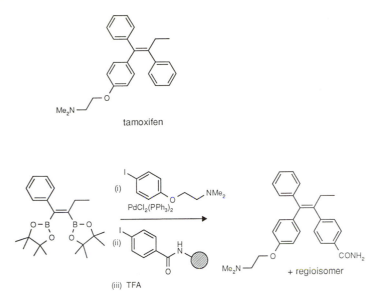

tamoxifen

Fig. 4.21   The use of resin capture in the synthesis of tamoxifen analogues.

multistep synthesis in one reaction vessel. The starting material would react with only one of the reagents, and once reaction was complete it could diffuse back into solution, now in a chemical state that makes it a substrate for the next reagent.

This principle has been applied for the three-step synthesis of a substituted pyrazole (Parlow 1995). The *sec*-phenethyl alcohol was mixed in cyclohexane with resin reagents that sequentially oxidized, brominated, and provided a hydroxypyrazole nucleophile (Fig. 4.22). The product was isolated in 48% overall yield that compared favourably with 42% yield for the stepwise sequence of reactions.

Although this methodology has yet to be applied to the synthesis of a combinatorial library, it is an obvious methodology to develop further especially in combination with the use of mixed reagents on ion exchange resin (Parlow 1996). This is likely to prove to be a rapid and highly efficient method for solution library synthesis.

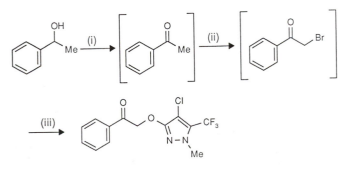

Fig. 4.22   A three-step synthetic scheme performed in one reaction vessel. The resins used were (i) poly(4-vinylpyridinium dichromate), (ii) perbromide on Amberlyst® A-26, and (iii) Amberlite® IRA-900 (4-chloro-1-methyl-5-(trifluoromethyl)-1H-pyrazol-3-ol).

## 12. Summary

Solution techniques clearly have their place in combinatorial chemistry. However, lacking the inherent advantages for rapid purification offered by solid phase synthesis, they are usually only applied when the synthetic routes are short, consisting of little more than one or two synthetic steps, and when the chemistry is likely to be reliable and high yielding. Consequently, solution synthesis is applied primarily to the synthesis of single compounds in parallel, often exploiting automated methods of solution dispensing, although there are the few examples of mixture synthesis in indexed libraries.

The future development of solution phase combinatorial chemistry lies rather in the further development of clever phase separation techniques. It has been noted that the advantages of solid phase chemistry lie in the ability to push chemical reactions to completion secure in the expectation that the excess reagents can be removed from the product attached to the solid phase by a simple filtration. Using excess reagents in solution synthesis is perfectly acceptable as long as there is a method of easy compound separation at the end of the chemical step. This chapter has described a number of techniques, several of them very new, that permit such easy phase separations. The use of fluorous reagents, soluble polymers, and resin capture demonstrate that it is no longer possible to clearly discriminate between solid phase and solution combinatorial techniques. Rather there is a continuum of methods that are currently being established, with the soluble polymer work appearing somewhere in the middle and the fluorous chemistry lying somewhat closer to the 'pure' solution phase chemistry end. It will be increasingly hard to pigeon-hole combinatorial techniques in the future—any approach that allows a ready, efficient, and successful synthesis of many compounds will be accepted as combinatorial chemistry if it is the most appropriate way of making large numbers of required compounds. It is very likely that some future library approaches may use hybrid phase techniques, where the synthesis commences in solution, key intermediates may be captured onto resin beads, and later cleaved back into solution for a fluorous reagent catalysed last step. The library components would follow a predetermined synthetic path through a range of different phases dictated by the reactions being performed, the reagents being used and the expected by-products and impurities. Everything is and will be tried—a natural consequence of the huge diversity of ideas in combinatorial chemistry.

## References

Bailey, N., Dean, A.W., Judd, D.B., Middlemiss, D., Storer, R., and Watson, S.P. (1996). A convenient procedure for the solution phase preparation of 2-aminothiazole combinatorial libraries. *Bioorg. Med. Chem. Lett.*, **6**, 1409–14.

Boger, D.L., Tarby, C.M., Myers, P.L., and Caporale, L.H. (1996). Generalized dipeptidomimetic template: solution phase parallel synthesis of combinatorial libraries. *J. Am. Chem. Soc.*, **118**, 2109–10.

Brown, S.D. and Armstrong, R.W. (1996). Synthesis of tetrasubstituted ethylenes on solid support via resin capture. *J. Am. Chem. Soc.*, **118**, 6331–2.

Carell, T., Wintner, E.A., Bashir-Hashemi, A., and Rebek, J. (1994*a*). A novel procedure for the synthesis of libraries containing small organic molecules. *Angew. Chem. Int. Ed. Engl.*, **33**, 2059–61.

Carell, T., Wintner, E.A., and Rebek, J. (1994*b*). A solution-phase screening procedure for the isolation of active compounds from a library of molecules. *Angew. Chem. Int. Ed. Engl.*, **33**, 2061–64.

Carell, T., Wintner, E.A., Sutherland, A.J., Rebek, J., Dunayevskiy, Y.M., and Vouros, P. (1995). New promise in combinatorial chemistry: synthesis, characterization, and screening of small-molecule libraries in solution. *Chem. Biol.*, **2**, 171–85.

Cheminat, A., Benezra, C., Farrall, M.J., and Frechet, J. (1980). Use of polymeric nucleophiles for the selective binding and removal of $\alpha$-methylene-$\gamma$-butyrolactone allergens from complex mixtures. *Tetrahedron Lett.*, **21**, 617–18.

Cheng, S., Comer, D.D., Williams, J.P., Myers, P.L., and Boger, D.L. (1996). Novel solution phase strategy for the synthesis of chemical libraries containing small organic molecules. *J. Am. Chem. Soc.*, **118**, 2567–73.

Curran, D.P. (1996). Combinatorial organic synthesis and phase separation: back to the future. *Chemtracts - Org Chem.*, **9**, 75–87.

Curran, D.P. and Hadida, S. (1996). Tris(2-(perfluorohexyl)ethyl)tin Hydride: A new fluorous reagent for use in traditional organic synthesis and liquid phase combinatorial synthesis. *J. Am. Chem. Soc.*, **118**, 2531–2.

Curran, D.P. and Hoshino, M. (1996). Stille couplings with fluorous tin reactants: attractive features for preparative organic synthesis and liquid-phase combinatorial synthesis. *J. Org. Chem.*, **61**, 6480–1.

Dooley, C.T. and Houghten, R.A. (1993). The use of positional scanning synthetic peptide combinatorial libraries for the rapid determination of opioid receptor ligands. *Life Sci.*, **52**, 1509–17.

Dunayevsky, Y.M., Vouros, P., Wintner, E.A., Shipps, G.W., Carell, T., and Rebek, J. (1996). Application of capillary electrophoresis–electrospray ionization mass spectrometry in the determination of molecular diversity. *Proc. Natl. Acad. Sci. USA*, **93**, 6152–7.

Gayo, L.M. and Suto, M.J. (1997). Ion-exchange resins for solution phase parallel synthesis of chemical libraries. *Tetrahedron Lett.*, **38**, 513–16.

Han, H., Wolfe, M.M., Brenner, S., and Janda, K.D. (1995). Liquid-phase combinatorial synthesis. *Proc. Natl. Acad. Sci. USA*, **92**, 6419–23.

Kaldor, S.W., Siegel, M.G., Fritz, J.E., Dressman, B.A., and Hahn, P.J. (1996*a*). Use of solid supported nucleophiles and electrophiles for the purification of non-peptide small molecule libraries. *Tetrahedron Lett.*, **37**, 7193–6.

Kaldor, S.W., Fritz, J.E., Tang, J., and McKinney, E.R. (1996*b*). Discovery of antirhinoviral leads by screening a combinatorial library of ureas prepared using covalent scavengers. *Bioorg. Med. Chem. Lett.*, **6**, 3041–4.

Keating, T.A. and Armstrong, R.W. (1996). Postcondensation modification of Ugi four-component condensation products: 1-isocyanocyclohexene as a convertible isocyanide. Mechanism of conversion, synthesis of diverse structures, and demonstration of resin capture. *J. Am. Chem. Soc.*, **118**, 2574–83.

Kim, R.M., Manna, M., Hutchins, S.M., Griffin, P.R., Yates, N.A., Bernick, A.M., *et al.* (1996). Dendrimer-supported combinatorial chemistry. *Proc. Natl. Acad. Sci. USA*, **93**, 10012–17.

Lawrence, K.B., Venepalli, B.R., Appell, K.C., Goswami, R., Logan, M.E., Tomczuk, B.E., *et al.* (1992). Synthesis and substance P antagonist activity of naphthimidazolium derivatives. *J. Med. Chem.*, **35**, 1273–9.

Parlow, J.J. (1995). Simultaneous multistep synthesis using polymeric reagents. *Tetrahedron Lett.*, **36**, 1395–6.

Parlow, J.J. (1996). The use of anion exchange resins for the synthesis of combinatorial libraries containing aryl and heteroaryl ethers. *Tetrahedron Lett.*, **37**, 5257–60.

Pirrung, M.C. and Chen, J. (1995). Preparation and screening against acetylcholinesterase of a non-peptide 'indexed' combinatorial library. *J. Am. Chem. Soc.*, **117**, 1240–5.

Pirrung, M.C., Chau, J.H-L., and Chen, J. (1995). Discovery of a novel tetrahydroacridine acetylcholinesterase inhibitor through an indexed combinatorial library. *Chem. Biol.*, **2**, 621–6.

Selway, C.N. and Terrett, N.K. (1996). Parallel-compound synthesis: methodology for accelerating drug discovery. *Bioorg. Med. Chem.*, **4**, 645–54.

Smith, P.W., Lai, J.Y.Q., Whittington, A.R., Cox, B., Houston, J.G., Stylli, C.H., *et al.* (1994). Synthesis and biological evaluation of a library containing potentially 1600 amides/esters. A strategy for rapid compound generation and screening. *Bioorg. Med. Chem. Lett.*, **4**, 2821–4.

Studer, A. and Curran, D.P. (1997). A strategic alternative to solid phase synthesis: preparation of a small isoxazoline library by 'fluorous synthesis'. *Tetrahedron*, **53**, 6681–96.

Studer, A., Hadida, S., Ferritto, R., Kim, S-Y., Jeger, P., Wipf, P., *et al.* (1997). Fluorous synthesis: a fluorous-phase strategy for improving separation efficiency in organic synthesis. *Science*, **275**, 823–6.

# 5 Encoded combinatorial synthesis

## 1. Introduction

One perennial problem in the testing of combinatorial libraries is the process of identifying the structure of the most active library component. As previous chapters have described, the amount of each compound produced combinatorially is frequently insufficient to permit routine chemical or spectroscopic structural analysis. However, the identity of the active component in a library can often be determined by its spatial position within an array, such as the microtitre block or the VLSIPS chip, and a knowledge of the chemistry performed in that position. Alternatively, the most potent compounds in a library can be identified by the process of recursive deconvolution. But there are occasions when the library is too large to permit rapid compound identification by spatial position or by iterative resynthesis and screening.

To solve the problems in identifying the active molecules from large libraries prepared on resin beads, scientists have devised various encoding methodologies. The basic principle of encoding is that every bead within the library has synthesized upon it a tagging compound or compounds that encode uniquely for the library compound on the bead (Fig. 5.1). In practise, given the ambiguity and uncertainty of organic synthesis, it is more accurate to state that the coding tags reveal the synthetic history of the bead to which they attach. Although this can refer unambiguously to one compound, the by-products or

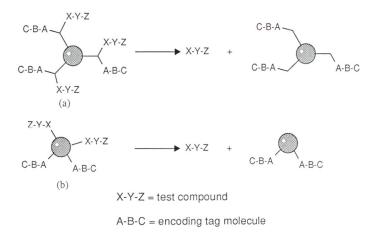

X-Y-Z = test compound

A-B-C = encoding tag molecule

**Fig. 5.1** The tag molecules (A–B–C) are synthesized on the bead in parallel with the test compounds (X–Y–Z). After library synthesis, the library compounds are cleaved for assay leaving the tags on the bead for analysis. Tags and library compounds may be (a) synthesized on a shared linker, or (b) could be synthesized on separate sites on the resin bead.

impurities that arise during the synthesis may also produce activity and thus might need to be identified following the decoding process. The encoding molecule structure is of a type that can be easily interpreted using spectroscopic or chromatographic techniques sensitive to the small amounts present. As the distinguishing properties of a specific set of tag molecules are often well characterized, these same molecules can be used repeatedly in different combinatorial libraries, in the certainty that unlike the molecules they encode, they will be consistent and predictable in their analytical properties.

Over the last few years, as the structures of library compounds have become more complex and distinct from the early biopolymers such as peptides or oligonucleotides, there has been a commensurate increase in encoding sophistication, such that decoding can now be undertaken in a rapid and unequivocal manner.

The libraries that benefit most from tagging strategies are those prepared using the mix and split paradigm, and destined to be screened in only one or at most a couple of screens. Ideally the beads are separated such that one compound at a time is assayed in solution. However, if portions of the compound on the bead can be removed sequentially, it is possible to test many compounds per well initially, and then focus in on the most potent compound by splitting the most active mixture into smaller and smaller sets of beads, until ultimately single compounds are tested (Fig. 5.2).

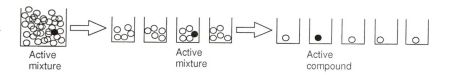

Active mixture      Active mixture      Active compound

**Fig. 5.2**  The identification of a single bead bearing an active component through iterative screening of successively smaller mixtures.

## 2.  Encoding requirements

For tagging to be an effective method for compound identification, the encoding method must meet several critical requirements:

(a)  The tag must be synthesized on the bead in parallel with the synthesis of the library component using chemistry that is mutually compatible. This obviously may limit the scope of synthetic chemistry available for the library, and careful consideration of the tagging scheme needs to be made to check that it will not adversely affect the compound synthesis.

(b)  The tag must be physically separable from the library compound so that screening of the library does not generate misleading biological data due to the presence of the encoding molecule. Additionally, the tag structure must be interpreted without interference from the ligand. Ideally, the test compound can be removed by a mild cleavage such as by the breaking of a photolabile linker, whilst the encoding molecule is left behind on the bead. Determination of the tag structure can either be done on the bead, or more vigorous or selective cleavage conditions could release the tag for chromatographic identification in solution.

(c)   The tag should be present in low concentrations so as not to occupy an unnecessarily large proportion of the resin bead functionality. As mentioned below, early tagging schemes required the synthesis of as much tag as library compound, but more subtle methods are now available that require the linking of just one tag molecule for every hundred or so library molecules.

(d)   The tag should be sequenceable by a rapid (possibly automated) spectroscopic or chromatographic technique. If this is not possible, the encoding scheme offers no advantage over the potentially complex structural analysis of the library component itself. Some of the most recent tagging schemes generate a mixture of encoding molecules that can be read via a chromatographic separation as a binary code that directly reveals the synthetic history of the bead.

## 3.   Examples of tagged libraries

Chemically encoded libraries fall into several main categories. As peptides and oligonucleotides are readily sequenced, the earliest tagged libraries were of peptides encoded by oligonucleotides (Section 3.1), peptides encoded by other peptides, and then non-sequenceable molecules encoded by peptides (Section 3.2). More recently the technology has moved on to use structurally novel encoding molecules that are read by chromatographic techniques (Sections 3.3 and 3.4).

### 3.1   Oligonucleotide tags

To date there have been several solutions to the combination of requirements listed in Section 2. In an early paper (Brenner and Lerner 1992) the concept of encoded combinatorial chemistry was first expounded. The authors suggested that an alternating parallel synthesis paradigm be employed to produce a library with the synthetic history encoded by 'genetic' oligonucleotide sequences. As each new monomer in the library synthesis is added to the solid phase, this would be mirrored by the addition of an oligonucleotide that specially encodes for the library monomer. As the library components are built up by routine mix and split, the oligonucleotide tags grow in parallel. Following biological assay, and identification of beads carrying the most potent compound, the polymerase chain reaction (PCR) could be used to amplify the encoding oligonucleotide on the bead, and this would be sequenced to reveal the structure of the library component.

This approach has been realized synthetically (Nielsen *et al.* 1993). A linker molecule was synthesized and attached to a CPG solid phase bead. The key branching point was a serine residue, providing side-arms for carrying both a peptide library component and the encoding oligonucleotide (Fig. 5.3)

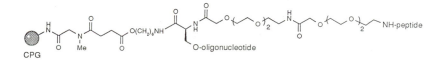
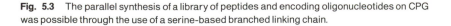

**Fig. 5.3**   The parallel synthesis of a library of peptides and encoding oligonucleotides on CPG was possible through the use of a serine-based branched linking chain.

leading to the synthesis of one encoding molecule for every library component.

Oligonucleotide tagging has been used by scientists at Affymax (Needels *et al.* 1993) for the deconvolution of a 823 543 ($7^7$) heptapeptide library, and these have been screened against an antidynorphin B monoclonal antibody. Using small (10 $\mu$m diameter) cross-linked polystyrene beads derivatized with a 1,12-diaminododecane linker gave a concentration of amine groups on the beads measured at 100 mmol $g^{-1}$ of resin, corresponding to a maximum peptide loading of 20 fmoles per bead. The amine-derivatized resin was reacted with an equal mixture of *N*–Fmoc–Thr(*t*-Bu)–OH (activated as the HOBt derivative) and 4-*O*-(dimethoxytrityl)oxybutyrate (activated as the succinimidyl ester), to provide an amino acid handle for building peptides and a protected hydroxyl for addition of oligonucleotides (Fig. 5.4).

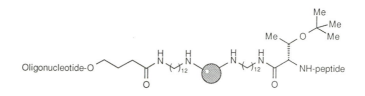

**Fig. 5.4** The Affymax linkers for parallel synthesis of a peptide and an oligonucleotide encoding chain.

The library was synthesized from seven L or D-amino acids on 35 mg of resin, leading to an estimated 200 beads containing each peptide sequence. The oligonucleotides were synthesized such that a unique dinucleotide represented each of the seven amino acids, and this encoding sequence of 14 bases was framed by PCR primer sequences giving a total length of 69 bases. The library compounds were assayed whilst still attached to the beads, and those that bound the antibody were detected by acquired fluorescence. A fluorescence-activated cell sorting (FACS) instrument was used to select the top 0.17% of the bead population with the brightest individual fluorescence. PCR amplification and sequencing of the oligonucleotide tags attached to these beads revealed the structures of the active peptides by reference to the planned encoding scheme. It was found that the peptide sequence TFRQFKV bound with the highest affinity to the antibody, a result consistent with the a previous observation that the preferred recognition sequence of the antibody is the hexapeptide RQFKVV.

This encoding scheme worked well because it is possible to store considerable information in the DNA template, and yet this can easily be read following PCR amplification. The tagging methodology combined with the fluorescence labelling of active peptide sequences and the rapid bead sorting achieved by the FACS instrumentation, provides an innovative method for library screening and compound structure determination.

## 3.2 Peptide tags

As with oligonucleotides, methods exist for the rapid sequencing of peptides. Several groups have attempted the encoded synthesis of library molecules using peptides as the tags. A group from Selectide and the Arizona Cancer

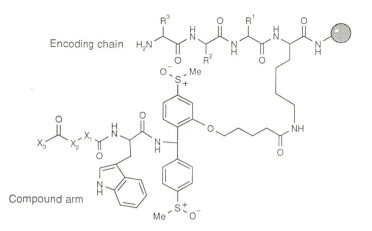

**Fig. 5.5** A lysine-based linker was employed to provide side-chains for the synthesis of ligand and encoding peptide. The non-sequenceable compound was attached through a safety-catch linker. It was necessary to reduce the two sulfoxides to sulfides before the carboxamide could be readily cleaved under acidic conditions.

Centre (Nikolaiev *et al.* 1993) described a library of non-peptides synthesized on solid phase in parallel with an encoding peptide chain. The key to this approach was a carefully designed branched linker that allowed the parallel synthesis of library and encoding molecule. The linker required a lysine residue to provide the branching point. The $\varepsilon$-amine provided attachment for the ligand, and the $\alpha$-amine was derivatized to build an encoding peptide sequence (Fig. 5.5). In one example of its use, a peptide library component was synthesized in parallel with an encoding peptide by using the differentiable base-labile Fmoc and acid-labile Boc protecting strategies for each arm. In another small library of 27 non-sequenceable components, only glycine, alanine, and leucine were used as encoding monomers in the peptide tag.

The non-peptide sequence was constructed by initially attaching an $\alpha$-bromo carboxylic acid to a tryptophan residue on the linker. This was reacted with either amines or an *N*-protected aminothiol to give amines or a thioether respectively, and the library compounds were finally capped with an acyl group. Cleavage from the resin gave products containing a C-terminal tryptophan carboxamide (Fig. 5.6)

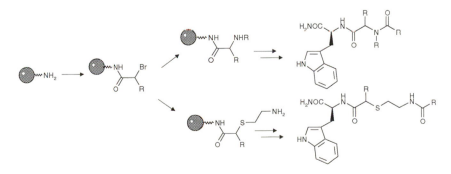

**Fig. 5.6** The synthesis of non-peptide compound sequences on the branched Selectide linker.

**Table 5.1**  The tagging scheme for a fictitious dipeptide library employing four distinct tag molecules

| Amino acid | Ala | Phe | Leu |
|---|---|---|---|
| Tag for position 1 | 1 | 2 | 1 and 2 |
| Tag for position 2 | 3 | 4 | 3 and 4 |

| Dipeptide | Tag | Dipeptide | Tag | Dipeptide | Tag |
|---|---|---|---|---|---|
| Ala–Ala | 1,3 | Ala–Phe | 1,4 | Ala–Leu | 1,3,4 |
| Phe–Phe | 2,4 | Phe–Ala | 2,3 | Phe–Leu | 2,3,4 |
| Leu–Leu | 1,2,3,4 | Leu–Ala | 1,2,3 | Leu–Phe | 1,2,4 |

for both their identity and their position in the dipeptide. Alanine in position 1 is encoded by tag 1, whereas in position 2 it is encoded by tag 3. Therefore if both tags 1 and 3 are observed by analysis the library molecule must be Ala–Ala. Leucine is encoded by two tag molecules—1 and 2 if in position 1, but 3 and 4 if in position 2 of the dipeptide. In all cases the tags observed are the combination of the specific encoding tag options for the constituent amino acids in specific positions in the dipeptide.

As it is the presence or absence of the tagging molecule that is important, just 20 tags will encode for $2^{20}-1$ or $1\,048\,575$ different possible library components. Furthermore, the tags are used at a level approximating to 0.5–1% of the quantity of the library compounds (one picomole total of all tags, in contrast to the 100 picomoles of library compound on each resin bead). Early libraries used a tag that attached to, and terminated the growing peptide chains of the library components. Although the presence of this tag stopped further growth of the peptide chain, the low level of loading did not introduce significant contamination of the final product. After synthesis, the peptides were assayed on the bead, and those beads that carried an active peptide sequence were identified through a colorimetric assay. Release of the tags by cleavage of a photolabile linker and identification of their structure by electron-capture capillary electrophoresis revealed the binary code that uniquely correlated with the structure of the ligand molecule.

Still *et al.* have used this method for the identification of peptide sequences that bind with high affinity to a synthetic receptor (Borchardt and Still 1994). A library of 50 625 ($15^4$) different acylated tripeptides was synthesized on resin beads. Each of the monomers (both amino acids and acyl groups) used in the library was represented by a combination of tag molecules. The 16 tag molecules employed were designated T1 to T16, where the number corresponds to their relative retention order by gas chromatography. The tags were assigned to the monomers as in Table 5.2, where the four digit binary code describes uniquely the amino acid or acyl group employed.

Before each of the coupling steps, the resin was reacted with a mixture of up to four tag molecules that uniquely represented the monomer about to be attached to the resin. The tagging reactions were followed by attachment of the

**Table 5.2** The four digit binary code tagging scheme that defines both the amino acid residue and its position in the library compound

R(C=O)–AA3–AA2–AA1–NH(CH$_2$)$_5$CONH–resin

| Vessel | N-terminal R | T13–T16 | AA3 | T9–T12 | AA2 | T5–T8 | AA1 | T1–T4 |
|---|---|---|---|---|---|---|---|---|
| 1 | Methyl | 1000 | Gly | 1000 | Gly | 1000 | Gly | 1000 |
| 2 | Ethyl | 0100 | D-Ala | 0100 | D-Ala | 0100 | D-Ala | 0100 |
| 3 | *iso*-Propyl | 0010 | L-Ala | 0010 | L-Ala | 0010 | L-Ala | 0010 |
| 4 | *tert*-Butyl | 0001 | D-Ser | 0001 | D-Ser | 0001 | D-Ser | 0001 |
| 5 | *tert*-Amyl | 1100 | L-Ser | 1100 | L-Ser | 1100 | L-Ser | 1100 |
| 6 | Trifluoromethyl | 0110 | D-Val | 0110 | D-Val | 0110 | D-Val | 0110 |
| 7 | *iso*-Butyl | 0011 | L-Val | 0011 | L-Val | 0011 | L-Val | 0011 |
| 8 | MeOCH$_2$ | 1001 | D-Pro | 1001 | D-Pro | 1001 | D-Pro | 1001 |
| 9 | Cyclopropyl | 1010 | L-Pro | 1010 | L-Pro | 1010 | L-Pro | 1010 |
| 10 | Cyclobutyl | 0101 | D-Asn | 0101 | D-Asn | 0101 | D-Asn | 0101 |
| 11 | Cyclopentyl | 1110 | L-Asn | 1110 | L-Asn | 1110 | L-Asn | 1110 |
| 12 | AcOCH$_2$ | 0111 | D-Gln | 0111 | D-Gln | 0111 | D-Gln | 0111 |
| 13 | Phenyl | 1011 | L-Gln | 1011 | L-Gln | 1011 | L-Gln | 1011 |
| 14 | Me$_2$N | 1101 | D-Lys | 1101 | D-Lys | 1101 | D-Lys | 1101 |
| 15 | Morpholino | 1111 | L-Lys | 1111 | L-Lys | 1111 | L-Lys | 1111 |

amino acid. The resin was mixed and split again into 15 pools before the tagging and coupling steps were repeated. The receptor, chemically labelled with a red dye, was mixed in solution with approximately one million beads carrying the library. Beads containing peptide sequences that bound strongly to the synthetic receptor could be identified by the development of a deep red colour. 50 beads with intense coloration were picked out and the encoding molecules were released by photolysis. The terminal hydroxyl groups in the tags were silylated with *bis*-trimethylsilylacetamide to aid their separation by capillary electrophoresis. Reading the CE traces as binary codes, the presence or absence of each of the 16 tags revealed the peptide sequences with the highest affinity for the receptors.

Later modifications of this encoding approach used rhodium catalysed carbene insertion chemistry to add the tags non-specifically to the cross-linked polystyrene backbone of the bead (Fig. 5.9) (Nestler *et al.* 1994). In each case very little tag molecule was actually used, and the inert nature of the tag did not compromise the synthesis of the library components.

To identify the active library components, two strategies have been used. First, the library was tested on the bead using a colorimetric assay that stained beads containing an active library component. The bead was selected and the compound identified by analysis of the photolytically cleaved tag molecules. Alternatively, the compounds were photolytically released into solution, and the tags were subsequently released, either by a ceric ammonium nitrate oxidation or by a more vigorous acid hydrolysis. In each case the tags were silylated and identified at subpicomolar levels by their retention times on electron-capture gas chromatography. The use of multiple tags again gave rise to a 'binary' code that could be read directly from the gas chromatogram by

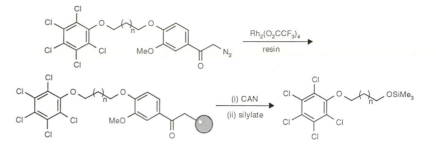

**Fig. 5.9** The haloaromatic tag molecules were added to the polystyrene backbone of the bead by a non-specific carbene insertion reaction. To identify individual compounds, the tag molecule was cleaved by a ceric ammonium nitrate (CAN) oxidation and silylated to facilitate gas chromatographic analysis.

the presence or absence of peaks, and the code uniquely defined the active library compound on the corresponding bead.

This tagging strategy has been used extensively by Still in the identification of the most potent components for synthetic receptors and in analysing novel synthetic receptor libraries for binders of specific peptides such as the enkephalins (Still 1996).

### 3.4   Secondary amine tags

Other robust chemical tagging schemes have been described. Amine tags have been incorporated into an *N*-[(dialkylcarbamoyl)methyl]glycine polymer attached to the resin bead at sites distinct from the library component (Ni *et al.* 1996). The tags themselves were a range of lipophilic secondary amines (e.g. HN(Et)($^n$Bu), HNMe($C_6H_{13}$), HN($C_8H_{17}$)$_2$, and HN($^n$Bu)$C_{12}H_{25}$), and these were introduced onto the TentaGel resin through the coupling of *N*-Boc-*N*-[(dialkylcarbamoyl)methyl]glycine monomers (Fig. 5.10). The encoding scheme relied on detecting the presence or absence of any of the dialkylamines used and was remarkably versatile. Thus a set of six tags could encode for $2^6-1$ (the 'null' tag combination was avoided) or 63 different compounds. If used in a three component synthesis, 18 tags (six for each of the three synthetic steps) could encode for up to 63 different components in each step,

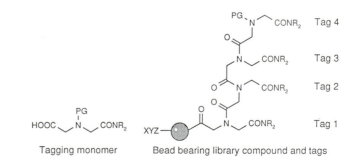

Tagging monomer        Bead bearing library compound and tags

**Fig. 5.10** The structure of the tagging monomer precursor, and the final derivatized bead bearing both the library compound X–Y–Z and the *N*-[(dialkylcarbamoyl)methyl]glycine 'oligomer' of tagging molecules.

or a total of $63^3$ (250 047) library compounds. This tagging procedure could be used in a routine split and mix library synthesis, with each tag or sets of tags being added either before or after the addition of each library compound monomer. As the amine tags were added as mixtures, the quantities of amines added were adjusted in inverse proportion to their reactivity. At the end of the synthesis, the library components were removed using either TFA or a photolytic cleavage, and the tags released by acid hydrolysis of the polyglycine with 6 N hydrochloric acid. The tags were derivatized with dansyl chloride and identified by reverse-phase HPLC. In an analogous fashion to the haloaromatic tags described in Section 3.3, the tags could be read as a binary code directly revealing the structure of the associated library compound.

This tagging method has been use to encode a library of 240 proline derivatives (Maclean *et al.* 1997). These compounds, prepared via a dipolar cycloaddition (see chapter 7 Section 3.4 for synthetic details), were tested individually in discrete wells for their ability to inhibit angiotensin converting enzyme (ACE). Such inhibitors have been developed and are currently marketed for their ability to reduce high blood pressure. Several active compounds were identified including an inhibitor with an $IC_{50}$ value of 610 pM. An interesting feature of this library synthesis is that it was a repeat of a library made previously by the same group. On the first occasion, the same compounds were made using a mix and split approach but without tagging, and the most active constituents had to be identified by an iterative deconvolution. In the more recent encoded library, the same optimum compound was rediscovered but this time a number of other active analogues were also identified. For example, the second and third most active compounds were discovered, but these were originally discarded in the iterative deconvolution approach. Clear SAR trends could be discovered by examining the most potent compounds—highly significant data for medicinal chemists intent on discovering a new drug molecule. This information is less readily forthcoming by iterative deconvolution, and would only be obtained by the resynthesis of individual compounds conjectured to be active based on the most active structure.

## 3.5   Isotope encoding

Tagging schemes work because the encoding molecules can be readily interpreted by chemical, chromatographic, or spectroscopic techniques. As mass spectrometry is amongst the most rapid and sensitive of analytical techniques, it is an obvious candidate for routine compound analysis. Indeed, many library syntheses use MS routinely for the analysis of individual library components (see Chapter 8, Section 4). Chemically homogeneous compounds generally appear as one signal: the molecular ion in the mass spectrum. If the compound contains an isotopic replacement, the mass ion will shift by a predictable amount. Using a predetermined mixture of isotopes in the compound will result in a characteristic MS ion pattern. This is the concept behind a method for library tagging using non-radioactive isotopes as the distinguishing feature (Geysen *et al.* 1996).

This principle has been used to propose four methods for the isotopic or mass encoding of bead libraries, and these methods could be used either alone or in combination to encode for any number of library compounds. In practise

**Table 5.3**   Ten possible epitope labelled dipeptide encoding molecules and their expected molecular ions by MS

| Dipeptide sequence | Molecular ion |
|---|---|
| Gly–Gly | |
| Gly–[$^{13}$C]Gly | +1 |
| [$^{13}$C]Gly–[$^{13}$C]Gly | +2 |
| [$^{13}$C]Gly–[$^{13}$C]$_2$Gly | +3 |
| [$^{13}$C]$_2$Gly–[$^{13}$C]$_2$Gly | +4 |
| Gly–Ala | +14 |
| Gly–[$^{13}$C]Ala | +15 |
| [$^{13}$C]$_2$Gly–Ala | +16 |
| [$^{13}$C]$_2$Gly–[$^{13}$C]Ala | +17 |
| [$^{13}$C]$_2$Gly–[$^{13}$C]$_4$Ala | +20 |

however it is always practical to minimize the size and consequently the synthetic effort required in introducing the encoding molecule. In its simplest incarnation, the isotope tagging approach uses a dipeptide encoding sequence made up from glycine and alanine monomers each containing various numbers of carbon-13 atoms. If natural glycine, $^{13}$C-glycine, $^{13}$C$_2$-glycine, natural alanine, $^{13}$C-alanine, and $^{13}$C$_4$-alanine are used, there are ten possible encoding dipeptides (Table 5.3). To maximize the sensitivity of the technique, the molecular ions are contained within a very narrow region of the mass spectrum. The presence of natural abundance of carbon-13 (1.1%) is not a problem in the interpretation of the mass spectra, as the relatively low molecular weight of the dipeptides ensures that the side peaks do not exceed 0.32 of the main peak intensity.

**Fig. 5.11**   The use of isotope labelled encoding dipeptides in the synthesis of the trimeric library, ABC.

It was proposed that the library compounds would be synthesized attached to the resin bead and released for screening in solution by cleavage of a selective linker (Fig. 5.11). If a particular compound was found to be active, the bead from which it came could be treated to cleave a second linker and release the encoding dipeptide. In a trimeric library constructed using the mix and split approach, the first monomer would be encoded by the isotopically labelled dipeptide. The final monomer would be defined by a knowledge of the pool from which the active bead was obtained. The central monomer remains to be identified, and this could be achieved by determining the mass of the active compound, and subtracting the masses of other two known monomers.

A more sophisticated expression of isotope labelling proposed the use of mixtures of labelled encoding monomer in each position of the encoding dipeptide. For example using natural glycine, $^{13}C$-glycine, and $^{13}C_2$-glycine in any equimolar combination in either position generates 25 possible different mass spectral 'bar codes', the position and shape of which is unique. Thus, the use of just three monomers can encode for 25 different library monomers.

This very recently described technique has been outlined in principle only. The simplicity of its use and the sensitivity and accuracy of MS suggests it could be widely applicable for future combinatorial libraries.

## 4. Radiofrequency tags

As molecular encoding can often throw up an incompatibility between the tagging chemistry and the chemistry of the ligand synthesis, there have consequently been a number of non-chemical tagging methods recently investigated. Amongst these, the concept of radiofrequency (RF) tagging has been simultaneously developed in a number of research laboratories.

RF encodable microchips inserted subcutaneously are widely used to tag and identify laboratory animals. These glass coated EEPROMs (electrically erasable, programmable read-only memories) can be encoded by remotely downloading RF binary information. Later the information can be retrieved by emitting RF signals that can be picked up over a distance of 75–150 mm. The recent chemical use of these chips was by enclosing them within a porous and inert container of resin beads. As these beads were subjected to a sequence of chemical transformations, the chip recorded a sequence of information that corresponded with the synthetic history of the beads. An alternative strategy relies on a set of pre-encoded read-only RF tags which are sorted prior to each synthetic transformation using the information in the encoded signal. With either approach, the chemical identity of the final compound can be read by reference to the RF code. Presumably, the transmission and reception of radio signals at low power levels will not interfere in any way with the synthesis of the compound library.

The use of these RF chips to label porous containers of resin beads is a modification of the Houghton 'tea-bag' approach to peptide synthesis, where by combining many beads in one container, they can all be simultaneously subjected to the same chemical transformation. The advance in this approach is that the chemical transformations applied may be related to the solid phase by referring to the sequence on the RF chip, which thus removes the considerable effort involved in manual tracking of reaction vessels.

In one example of the use of Radiofrequency Encoded Combinatorial (REC$^{TM}$) chemistry, a library of 24 peptides was synthesized using a $3 \times 2 \times 2 \times 2$ mix and split paradigm (Nicolaou *et al.* 1995). 96 microreactors were constructed from an inert porous support containing TentaGel resin beads and a semiconductor RF tag. The reactors were initially split into three pools. Each pool was addressed by write encoding with a unique RF signal before coupling onto the resin an Fmoc-protected amino acid encrypted by the radio code. The reactors were pooled and split into two groups followed by unique RF encoding and the coupling of a second monomer. The synthetic process was completed to generate tetrapeptide sequences on the resin, whilst in parallel the microchips were encoded with information that mapped the precise sequence of synthetic steps to which the microreactor had been subjected. The memory capacity of the chips is such that it is possible to record details of the reaction conditions and all the steps of a multistep synthetic sequence.

Following synthesis, the peptides were cleaved from the resin and the chip could be interrogated to define the identity of the peptide released from each microreactor. The structures were confirmed by both mass spectrometry and ${}^1$H NMR, and shown to be between 67% and 97% pure by HPLC.

A second group has used a similar process to record the synthesis of a combinatorial library of protein tyrosine phosphatase PTP1B inhibitors (Moran *et al.* 1995). A tripeptide acylated at the N-terminus with *p*-chlorocinnamic acid was the template for the library and 125 analogues (**5.1**) were prepared on Rink resin in a 5 × 5 × 5 mix and split process. The resin beads were enclosed in a polypropylene capsule containing the RF encodable microchip, and as with the example of Nicolaou *et al.* each step in the synthetic sequence was recorded on the microchip by the unique binary code. At the end of the synthesis the individual capsules were scanned and distributed into the 96 wells of a solid phase reaction block system and the derivatized peptides cleaved from the resin. Thus the identity of each peptide in each well had been determined and electrospray mass spectrometry confirmed the presence of all 125 expected products.

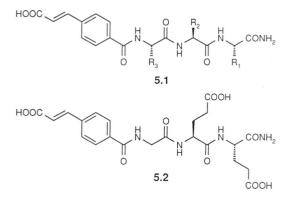

Screening of the peptides against PTP1B revealed several inhibitors, and purified samples revealed that the cinnamyl derivative of Gly–Glu–GluNH$_2$ (**5.2**) had an IC$_{50}$ value of 44 nM.

In summary, the RF addressable microchips offer a novel approach to the encoding of synthetic information that can unequivocally identify a compound following library synthesis and screening. Because of their glass coating they are chemically inert to nearly every possible chemical process as well as having a wide range of temperature tolerance. The process of encoding information is totally orthogonal to the synthesis, as the chemistry will not affect the information on the microchip, and the RF encoding will not affect the molecules being synthesized on the solid phase. The real value of this approach will be realized when RF encoding is used for large libraries where the number of reactor vessels employed is too large to permit tracking by other methods. There will be a number of logistical difficulties to overcome in handling large numbers of microreactors and automated approaches to directed sorting are now being used by the groups active in the area. Finally, unlike the chemical tagging methods described in detail earlier in this chapter, where the units involved in synthesis were individual beads, in these RF tagging approaches, the synthesis unit is the microreactor. The size of the microchip defines the minimum size of these microreactors, and there is probably a minimum size limit for the chip, and a lower limit for the amount of resin than can consequently be used for the synthesis of each library compound.

## 5. Laser optical encoding

A huge variety of encoding strategies have been explored for tagging combinatorial library components. Amongst the many coding techniques that are non-invasive, and therefore cannot affect the library chemistry is the use of laser bar code etching (Xiao *et al.* 1997). A readable two-dimensional bar code is generated by laser-etching with a carbon dioxide laser onto an inert alumina ceramic plate. The laser optical synthesis chip (LOSC) is a polypropylene square 10 mm × 10 mm which has been radiolytically grafted with cross-linked polystyrene to provide a synthesis support for up to 8 $\mu$mol of a single compound. In the centre of the LOSC is the 3 mm × 3 mm ceramic bar code which is encoded prior to the commencement of synthesis.

A 'directed sorting' strategy was used for the synthesis of a library of 27 oligonucleotides. In the directed sorting approach, there is zero redundancy amongst the solid supports. This means that there is one chip for each compound in the library. The library is prepared by first scanning the codes on the chips and sorting the chips prior to each synthetic step. At the end of the synthesis the structure of the compound on each LOSC can be determined by reference to the code. The LOSC can be varied in shape and size to meet the needs of the synthesis. The lower limit of size is dictated by the bar code which can be as small as 0.5 mm and still camera readable (Fig. 5.12).

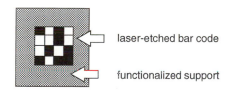

laser-etched bar code

functionalized support

**Fig. 5.12** The construction of the LOSC support for laser encoded library synthesis.

This approach offers yet another method of determining the structure of compounds in a combinatorial library, and offers the possibility of being streamlined through automatic handling of the LOSC supports.

## 6. Fluorophore encoding

One final method of library encoding is little more than a concept at present but preliminary experiments have been described in a few papers. The first study in this area described the use of labelled TentaGel beads bearing either fluorescein (0.01%) or erythrosin (0.02%) dyes (Egner *et al.* 1997). The use of a confocal fluorescence microscope to examine a single bead permitted the fluorescence spectrum to be recorded. Keeping the loading below 0.1%

prevented dye saturation and maintained a linear relationship between loading and fluorescence. Beads that were tagged simultaneously with two dyes at 0.01% showed that some internal quenching was taking place but was incomplete, and the identity of the two dyes could still be ascertained.

This group proposed that tagging beads with a single dye could correspond to the residue added to the bead in the first synthetic step of combinatorial library synthesis. In this way the analysis of the active component of a mix and split library could be greatly facilitated.

A second paper has described similar studies (Scott and Balasubramanian 1997). This group also found that the spectroscopic properties of fluorescent molecules attached to TentaGel and polystyrene resin beads at low levels are very similar to their properties in solution. The excitation and emission frequency bands are not shifted or broadened significantly and the fluorescence intensity was found to be proportional to the quantity of the fluorophore on the bead.

Thus the predictability of fluorophore behaviour allowed them to be used in a diagnostic fashion to label a resin bead. It was again demonstrated that several different fluorophores could be attached simultaneously at very low loading levels to the same resin bead. As both excitation and emission wavelengths can be varied, overlapping rarely occurred and all four fluorophores could be readily detected on the bead without quenching being observed. With the additional possibility of varying the quantity on the bead, a tagging scheme could be constructed where the presence or absence of fluorophores and their quantification could be used as a record of the synthetic history of a particular bead. It has been suggested that the beads should be pre-encoded with the fluorophores and the bead sorted by a FACS instrument prior to each library synthesis step.

These methods have yet to be realized in the synthesis of a combinatorial library but the approach offers a suitable alternative to other tagging schemes. Fluorophore tagging provides the advantage of high sensitivity and avoids any need to cleave the encoding molecules from the bead before decoding. However, the potential chemical and photochemical sensitivity of the tag molecules are problems that would need to be circumvented before this could be a generic and reliable encoding scheme.

## 7.   Summary

This chapter has described a diverse range of chemical and physical encoding methods explored by combinatorial chemists to monitor the synthesis of library compounds. Each of these methods has advantages and some also have evident problems or disadvantages. Thus the chemist wishing to make large libraries needs to decide which method, if any, is most appropriate for their needs, as usually the type of library, scale of synthesis, and ultimate use will dictate the preferred tagging scheme. Clearly, many of these approaches are highly specialized, requiring sophisticated equipment and considerable technology development to use effectively. Fortunately, for many of these methods, access to the technology is possible through collaboration with the biotechnology companies that have developed these methods and hold the relevant intellectual property rights.

# References

Borchardt, A. and Still, W.C. (1994). Synthetic receptor binding elucidated with an encoded combinatorial library. *J. Am. Chem. Soc.,* **116**, 373–4.

Brenner, S. and Lerner, R.A. (1992). Encoded combinatorial chemistry. *Proc. Natl. Acad. Sci. USA*, **89**, 5381–3.

Egner, B.J., Rana, S., Smith, H., Bouloc, N., Frey, J.G., Brocklesby, W.S., *et al.* (1997). Tagging in combinatorial chemistry: the use of coloured and fluorescent beads. *J. Chem. Soc. Chem. Commun.*

Geysen, H.M., Wagner, C.D., Bodnar, W.M., Markworth, C.J., Parke, G.J., Schoenen, F.J., *et al.* (1996). Isotope or mass encoding of combinatorial libraries. *Chem. Biol.*, **3**, 679–88.

Kerr, J.M., Banville, S.C., and Zuckermann, R.N. (1993). Encoded combinatorial peptide libraries containing non-natural amino acids. *J. Am. Chem. Soc.*, **115**, 2529–31.

Maclean, D., Schullek, J.R., Murphy, M.M., Ni, Z-J., Gordon, E.M., and Gallop, M.A. (1997). Encoded combinatorial chemistry: synthesis and screening of a library of highly functionalised pyrrolidines. *Proc. Natl. Acad. Sci. USA*, **94**, 2805–10.

Moran, E.J., Sarshar, S., Cargill, J.F., Shahbaz, M.M., Lio, A., Mjalli, A.M.M., *et al.* (1995). Radio frequency tag encoded combinatorial library method for the discovery of tripeptide-substituted cinnamic acid inhibitors of the protein tyrosine phosphatase PTP1B. *J. Am. Chem. Soc.*, **117**, 10787–8.

Needels, M.C., Jones, D.G., Tate, E.H., Heinkel, G.L., Kochersperger, L.M., Dower, W.J., *et al.* (1993). Generation and screening of an oligonucleotide-encoded synthetic peptide library. *Proc. Natl. Acad. Sci. USA*, **90**, 10700–4.

Nestler, H. P., Bartlett, P.A., and Still, W.C. (1994). A general method for molecular tagging of encoded combinatorial chemistry libraries. *J. Org. Chem.*, **59**, 4723–4.

Ni, Z-J., Maclean, D., Holmes, C.P., Murphy, M.M., Ruhland, B., Jacobs, J.W., *et al.* (1996). Versatile approach to encoding combinatorial organic syntheses using chemically robust secondary amine tags. *J. Med. Chem.*, **39**, 1601–8.

Nicolaou, K.C., Xiao, X-Y., Parandoosh, Z., Senyei, A., and Nova, M.P. (1995). Radiofrequency encoded combinatorial chemistry. *Angew. Chem. Int. Ed. Engl.*, **34**, 2289–91.

Nielsen, J., Brenner, S., and Janda, K.D. (1993). Synthetic methods for the implementation of encoded combinatorial chemistry. *J. Am. Chem. Soc.*, **115**, 9812–13.

Nikolaiev, V., Stierandova, A., Krchnak, V., Seligmann, B., Lam, K.S., Salmon, S.E., *et al.* (1993). Peptide-encoding for structure determination of nonsequenceable polymers within libraries synthesized and tested on solid phase supports. *Pept. Res.*, **6**, 161–70.

Ohlmeyer, M.H.J., Swanson, R.N., Dillard, L.W., Reader, J.C., Asouline, G., Kobayashi, R., *et al.* (1993). Complex synthetic chemical libraries indexed with molecular tags. *Proc. Natl. Acad. Sci. USA*, **90**, 10922–6.

Patek, M. and Lebl, M. (1991). Safety-catch anchoring linkage for synthesis of peptide amides by Boc/Fmoc strategy. *Tetrahedron Lett.*, **32**, 3891–4.

Scott, R.H. and Balasubramanian, S. (1997). Properties of fluorophores on solid phase resins; implications for screening, encoding and reaction monitoring. *Bioorg. Med. Chem. Lett.*, **7**, 1567–72.

Still, W.C. (1996). Discovery of sequence-selective peptide binding by synthetic receptors using encoded combinatorial libraries. *Acc. Chem. Res.*, **29**, 155–63.

Vágner, J., Barany, G., Lam, K.S., Krchnák, V., Sepetov, N.F., Ostrem, J.A., *et al.* (1996). Enzyme-mediated spatial segregation on individual polymeric support beads: Application to generation and screening of encoded combinatorial libraries. *Proc. Natl. Acad. Sci. USA*, **93**, 8194–9.

Xiao, X-Y., Zhao, C., Potash, H., and Nova, M.P. (1997). Combinatorial chemistry with laser optical encoding. *Angew. Chem. Int. Ed. Engl.*, **36**, 780–2.

# 6 Beyond peptide libraries

## 1. The development of non-peptide libraries

Chapters 2 and 3 described the explosive growth in methods for the rapid synthesis of peptide libraries, primarily for immunological studies, reflecting the critical role played by solid phase peptide chemists in the initial development of combinatorial chemistry methods. Following the seminal 1991 *Nature* papers from Houghten and Lam, there was an enormous growth in interest in combinatorial chemistry from the entire chemical community in general, and from pharmaceutical chemists in particular. The technology clearly had the potential to accelerate the processes involved in drug discovery, and many pharmaceutical and specialized biotechnology companies began their own investigation of combinatorial methods. Whilst most were based on solid phase methods, primarily resin beads, there was an enormous diversity in the approaches used to generate and test compounds. In addition, the drive was to apply rapid chemical methods to non-peptide structures, as these were generally accepted to be superior starting points for drug discovery programmes. Peptides are poorly absorbed if administered orally, and are subject to rapid metabolism and elimination if administered systemically (intravenously for example).

This chapter focuses initially on the use of combinatorial chemistry to prepare libraries of the other natural oligomers; oligonucleotides and oligosaccharides. There follows a description of other oligomeric materials designed to mimic the properties of peptides. Towards the end of the chapter I describe a number of library approaches to the synthesis of compounds that are barely peptide-like at all. Pharmaceutical chemists in particular have made great strides in compound library design in order to get away from peptides and prepare compounds that are closer to the typical non-oligomeric molecules generally developed as drugs. This chapter tracks the evolution of chemistry away from the naturally occurring oligomers and towards the library synthesis of diverse low molecular weight organic molecules.

## 2. Oligonucleotide libraries

Like peptides, oligonucleotides are polymers that can be synthesized by consistent and reliable coupling conditions, and thus had been the focus of solid phase synthesis before the advent of combinatorial methodologies. Furthermore, because it is possible to apply PCR to amplify exceedingly small amounts of DNA, it is possible to make very large libraries (up to $10^{15}$ different sequences) in the expectation that isolated active molecules can be identified by amplification and automated sequencing. The techniques for high-throughput synthesis can readily be applied to prepare libraries of oligonucleotides for a variety of applications. In Chapter 3, Section 3.1, the marriage between oligonucleotide synthesis and novel planar solid phases was described as leading to sophisticated methods for DNA sequencing and analysis, but library synthesis on beads has also been used in the search for pharmacologically active molecules.

A library of phosphorothioate oligonucleotides containing all possible eight nucleotide sequences was synthesized on controlled pore glass beads (Wyatt *et al.* 1994). The library was constructed in 16 mixtures of 4096 oligomers of the structure NNXNXNNN, where N was a mixture of the four bases and X was a defined position. As there are only four monomers used in oligonucleotide synthesis, and the coupling reactivity is largely predictable, it is possible to circumvent the mix and split synthesis procedure, and mixtures can be introduced into the library by subjecting support-bound precursors to a mixture of activated monomers (amidites) previously shown to introduce equal quantities of nucleotide bases.

The library was screened in an acute HIV-1 infection cell-based assay, and the $IC_{50}$ values of active mixtures ascertained. Using a selection strategy described by Isis scientists as SURF (synthetic unrandomization of randomized fragments), the most active single oligonucleotide sequence was determined to be TTGGGGTT, with an $IC_{50}$ value of 0.3 $\mu$M. It was found furthermore that this oligomeric sequence readily formed multimeric complexes, one of which, a tetrameric form, was responsible for the antiviral activity. The tetramer was found to exert its anti-HIV activity through binding to the viral envelope protein gp120 and inhibiting both cell-to-cell and virus-to-cell infection.

The same group has also constructed combinatorial oligonucleotide libraries using a variety of modified and natural nucleosides (Davis *et al.* 1995). The pentamers synthesized had the generic structure XNNNT where X was a defined position, N was a mixture of all 12 nucleosides, and T corresponded to a thymidine-derivatized controlled pore glass solid support. The first round of synthesis generated 12 subsets each containing 1728 compounds ($12^4$) which were tested against two anti-inflammatory targets: the enzyme phospholipase $A_2$ and the leukotriene $B_4$ receptor. Iterative deconvolution led to the identification of unique micromolar inhibitors of these two targets.

From the studies of the Isis group it was concluded that due to errors in synthesis and screening, positional scanning and iterative deconvolution are both likely to fail in discovering the most potent compound in a library. This has been circumvented by the development of 'Mutational SURF', a method that independently synthesizes and tests 'mutants' related to the active selected sequences and optimises the structure and activity of the initial active leads (Freier *et al.* 1996).

Libraries of oligonucleotides have also been used by other groups in the search for therapeutic agents. For example, a process called SELEX (systematic evolution of ligands by exponential enrichment) has been developed to find inhibitors of various targets including HIV-1 reverse transcriptase (Schneider *et al.* 1995). The combinatorial chemistry component involved the random synthesis of single-strand DNA oligonucleotide libraries containing 35 random nucleotide positions flanked by invariant sequences important for later amplification by the PCR. The library was incubated with the HIV-1 reverse transcriptase (RT) enzyme and single-strand DNA sequences that bound were separated by using a size-exclusion chromatography step. Amplification of the sequences that bound to HIV-1 RT was followed by further incubation steps. After several rounds of this selection process, highly potent inhibitors of the enzyme ($K_i$ values as low as 0.3 nM) had been discovered.

Whilst a sophisticated use of DNA technology to find novel therapeutic agents, the combinatorial chemical component of SELEX is of limited interest to synthetic chemists, and the technique will not be described here further.

## 3.  Oligosaccharides

Oligosaccharides are important mediators for a large number of biological processes, such as cell–cell recognition and adhesion, hormone–cell recognition, and viral or bacterial adhesion to host cells. As such they provide significant target molecules for pharmacological intervention, and various disease states and infections might be controlled by the administration of molecules that mimic endogenous oligosaccharides or can block the recognition or adhesion processes. Consequently it might be considered that oligosaccharides are prime candidates for synthesis through combinatorial chemistry, and that enhancing the molecular diversity of these compounds might provide novel and pharmacologically significant agents. However the number of libraries that have been prepared so far is exceedingly limited (Sofia 1996).

The main explanation for this puzzling lack of exploitation lies in the complexity presented by the monosaccharide building blocks. Unlike amino acids, that in the main have only one amine and one carboxylic acid that can be involved in the formation of peptide bonds, monosaccharide sugars have a multitude of hydroxyl groups each of which can be glycosylated by another sugar molecule, presenting a range of regiochemical and stereochemical possibilities. To simplify synthesis there needs to be a reliable protecting group strategy. Furthermore, there are many different ways of forming the glycosidic link, none of which provide a generic solution to the varied needs of oligosaccharide synthesis. The stereochemistry of the glycosidic bond can either be axial or equatorial, and there may need to be control over the production of linear or branched chains. As an illustration of the potential complexity of oligosaccharides, we can consider the diversity possible from just the nine common monosaccharides in mammalian biosynthesis. All of these are present in their pyranose forms and can have either three or four hydroxyl groups that may undergo glycosylation. This diversity leads to 119 736 possible trisaccharides and over 18 million tetrasaccharides! Such massive structural diversity is the dream of combinatorial chemists, but this could so readily become a nightmare if the regio- and stereochemistry cannot be tightly controlled.

Unfortunately, there are very few good methods for the solid phase synthesis oligosaccharides in a high and consistent yield. Furthermore, having made and screened a library, unlike peptides or oligonucleotides, there are no simple methods for the analysis or sequencing of the sugar moieties. Attempts to produce oligosaccharides on solid phase actually pre-date the advent of combinatorial chemistry, but throughout this work the main focus has been the development of reliable glycosylation strategy. A preferred approach would be to use a single form of glycosyl donor that is effective irrespective of the nature of the glycosidic linkage being formed.

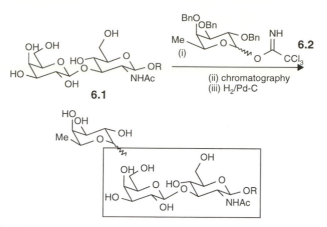

**Fig. 6.1**  The random glycosylation of the disaccharide, $\beta$Gal(1→3)$\beta$GlcNAc (**6.1**) giving a mixture of six trisaccharide products. R = (CH$_2$)$_8$OC$_6$H$_4$$p$OMe, a chain that was used for both ease of isolation and to enhance UV detection.

One attempt to overcome the glycosidic link problem has been to use a totally random approach to oligosaccharide synthesis (Kanie *et al.* 1995). The objective of this study was to produce mixtures of trisaccharides from the fucosylation of a disaccharide and test the mixtures in a biological assay. $\beta$Gal(1 → 3)$\beta$GlcNAc (**6.1**) was one of the substrates employed in the random glycosylation and this disaccharide possesses six hydroxyls, each of which would need to be derivatized to make all possible trisaccharides. To make these six compounds in a controlled synthetic fashion requires extensive protection and deprotection chemistry strategies for each of the disaccharide precursors with chromatographic purifications at essentially every step of the process.

In contrast the random glycosylation process took the disaccharide in an unprotected form and reacted it with two equivalents of a fully protected tri-*O*-benzylfucopyranosyl trichloroacetimidate (**6.2**) in the presence of BF$_3$–Et$_2$O as a promoter. The product mixture contained around 30% of the desired trisaccharide products and these were separated using reverse-phase chromatography from starting disaccharides containing no benzyl groups or tetrasaccharides containing six benzyl groups. Finally the trisaccharide fraction was hydrogenated to remove the benzyl protecting groups (Fig. 6.1).

The individual isomers were separated by repeated chromatography at both the protected and deprotected stages and it was found that all six products had been formed. The distribution of $\alpha$-fucosylation was 12% $\alpha$(1 → 4), 22% $\alpha$(1 → 6), 19% $\alpha$(1 → 2'), 23% $\alpha$(1 → 3'), 8% $\alpha$(1 → 4'), and 16% $\alpha$(1 → 6'), rather than the statistical distribution of 17% of each isomer. The random glycosylation was applied to the fucosylation of $\beta$Gal(1 → 4)$\beta$GlcNAc and $\beta$Gal(1 → 6)$\beta$GlcNAc with similar success, thus demonstrating this to be an effective way rapidly to construct small libraries of oligosaccharides.

The same group has used random glycosylation for the synthesis of disaccharides (Ding *et al.* 1996). Unprotected *N*-acetylglucosamine (**6.3**) was glycosylated by fully protected peracetylated galactopyranosyl imidate in dioxane to give a mixture of all six $\alpha$- and $\beta$-disaccharides (Fig. 6.2).

It was found that the ratio of the products could be varied by the use of different glycosyl donors. The anomeric configuration of the products could

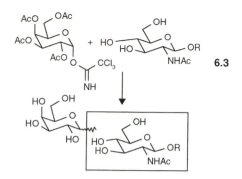

**Fig. 6.2** The random glycosylation of *N*-acetylglucosamine (**6.3**).

be biased towards the α-glycosides by the use of tetrabenzylated protected imidate (**6.4**) or phosphite (**6.5**), whereas all the six products could be obtained if the tetraacetylated imidate (**6.6**) or phosphite (**6.7**) are used.

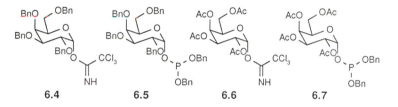

The next advance in this technique will be to demonstrate whether it can be used for other glycosylating sugars, and also to see whether multistep reactions can be performed to extend the size and the complexity of the oligosaccharide products.

Molecular diversity in an oligosaccharide library will result from a large range of available glycosyl donors and acceptors. A method for the use of the same building blocks as both donors and acceptors for a glycosylation has recently been reported (Boons *et al.* 1996). Using this latent-active glycosylation approach, a common building block can be converted to a glycosyl acceptor by cleavage of an acetyl protected hydroxyl, or by isomerization of an anomeric allyl ether group, into a glycosyl donor (Fig. 6.3). Trimethylsilyl triflate catalysed glycosylation effectively couples the hydroxyl saccharide with the anomeric vinyl ether. As the allyl/vinyl protocol is a valuable method for extending the versatility of monosaccharides, just four building blocks could be used to generate 32 disaccharides as 16 mixtures of anomers. If these products are mixed, and split into four pools, a further glycosylation will generate four mixtures of 64 trisaccharides.

As an illustration of this approach, a small library of 20 trisaccharides have been produced. As the chemistry was done in solution, and an excess of reagents was required to optimize yields, size-exclusion chromatography was required to purify the intermediate disaccharides. Additionally, it was noted that the glycosylation reaction was not equally successful in every case, although hydrolysis of the products to monosaccharides for analytical HPLC showed that glucose, galactose, and fucose were present in approximately the expected ratio.

As this example demonstrates, the difficulties of producing a combinatorial library of oligosaccharides have been created by the unavailability of generic

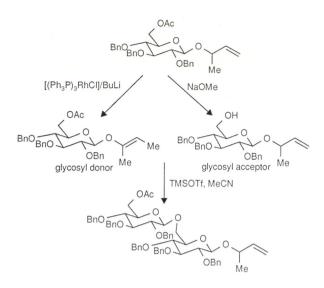

**Fig. 6.3** The use of a common building block to produce a disaccharide by conversion to both a glycosyl donor and acceptor.

glycosylation conditions that work irrespective of the nature of the glycosylating agent and the glycosyl acceptor. This situation has been ameliorated recently by the development of anomeric sulfoxides for glycosylation (Yan *et al.* 1994). Although a *β*-(1-6)-linked trisaccharide could readily be formed on solid phase by glycosylation of a primary alcohol, the real challenge came in attempting to glycosylate a secondary alcohol. It was found that by attaching the acceptor, the *p*-hydroxythiophenyl glycoside of 2-azido-2-deoxy-4,6-*O*-benylidene-D-glucose to Merrifield resin, the sulfoxide glycosylation reaction could be used to give both *α* and *β* glycosidic linkages to secondary alcohols stereospecifically and in high yield (Fig. 6.4).

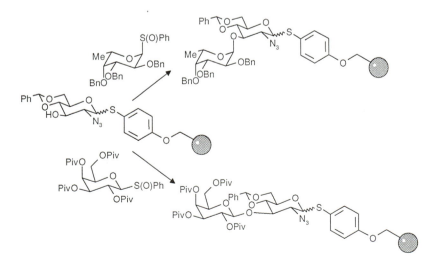

**Fig. 6.4** The stereospecific glycosylation of a resin-bound secondary hydroxyl acceptor with anomeric sulfoxides.

The reaction proceeded under mild conditions; the sulfoxide being activated by the addition of triflic anhydride, and cleavage from the resin was effected by mercuric trifluoroacetate to give the disaccharides as the only detectable sugars. The yields of 85–96% for these glycosylations are higher than any corresponding solution phase reaction, reflecting the advantage of repeating the reaction on solid phase to raise the yields of difficult couplings.

The paper closed with the comment that with yields for the glycosylation reaction approaching those obtained in solid phase peptide and oligonucleotide reactions, a practical strategy for the chemical synthesis of oligosaccharides was now in sight. The use of these reactions for the first true combinatorial library synthesis was reported in a seminal paper in the journal *Science* (Liang *et al.* 1996). The significance of this piece of work lies in the way that so many of the key elements of combinatorial chemistry are brought together to support the remarkable oligosaccharide chemistry. In addition to the use of the anomeric sulfoxide glycosyl donors to prepare a library of 1300 compounds, the library was prepared on beads using the haloaromatic tagging scheme devised by Still (see Chapter 5, Section 3.3), and in screening the compounds, novel and potent oligosaccharide ligands for a lectin protein were discovered.

The compounds synthesized were constructed from three components (Fig. 6.5). First, a set of six glycosyl acceptors were individually attached to TentaGel amine resin using HBTU (2-(benzotriazol-1-yl)-1,1,3,3-tetramethyl-uronium hexafluorophosphate) coupling conditions. Following hydrazine catalysed deprotection of an acetate-protected alcohol in the glycosyl acceptor, the resin portions were mixed and redivided. The second step was the glycosylation with each of a set of 12 anomeric sulfoxides using the mild triflic anhydride catalysed method. Following another mix and split step, the azide was reduced with trimethyl phosphine, and the product amine was acylated with a diverse range of 18 acylating reagents, including acyl chlorides, anhydrides, sulfonyl chlorides, isocyanates, and diketene. Resin portions were also retained as the azide and the amine to increase the overall size of the library. To allow the identification of the oligosaccharides on each bead, the resin portions were encoded with chemical tags after each chemical step. The final library was deprotected by the sequential treatment with TFA and lithium hydroxide.

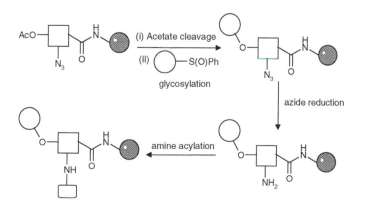

**Fig. 6.5** The strategy employed for a combinatorial synthesis of a capped disaccharide library.

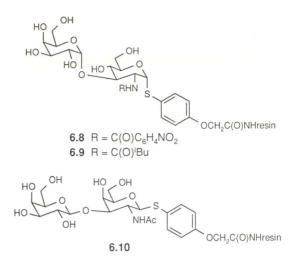

**6.8**  R = C(O)C$_6$H$_4$NO$_2$
**6.9**  R = C(O)$^i$Bu

**6.10**

The library was synthesized in a direction that ensured that each step could be forced to completion by the addition of excess glycosyl donor or acylating agent. To avoid complications in the biological screening, it was essential that each bead contained only one compound. The screening of the library against the lectin from *Bauhinia purpurea* took place whilst the compounds were still attached to the resin beads. The use of a colorimetric assay highlighted beads containing an active structure, and the tag molecules were decoded to reveal the active oligosaccharide structures. It was found that two derivatives of Gal-α-1,3-GlcNR-α-thiophenyl glycoside (**6.8** and **6.9**) when attached to the TentaGel beads were found to be even better ligands for the lectin than the known ligand (**6.10**) on resin. The new compounds were resynthesized and tested in solution and were found to retain their affinity for the lectin.

Oligosaccharide combinatorial chemistry has made a slow start, but as described in this section, this is an expected consequence of the restricted number of techniques available for reliable glycosylation reactions. However, as the search for better synthetic methods is a focus of considerable current interest for both combinatorial and carbohydrate chemists, the development of oligosaccharide libraries will inevitably be a growth area.

## 4.  Other oligomers

There have been many attempts to express the key features of a peptide structure in non-peptide oligomers. Whilst this may seem somewhat contradictory, it is possible to retain the side-chains in a peptide-like structure whilst replacing the vulnerable peptide linked backbone. In particular, it may be important to replace the amide bond with a group that favours the preferred active conformations expressed by the natural peptides, so that the side-chains may be spatially arranged to interact with receptor or enzyme subsites. This is an area where the advent of combinatorial techniques have had a subtle but increasingly important influence. Even before the advent of combinatorial chemistry, chemists wished to express the activity of a peptide in a non-peptidic structure, and whilst there were and still are no reliable rules for this procedure, there have been a multitude of proposals for peptide mimics. As library methods have come to the forefront, peptidomimetic designs and

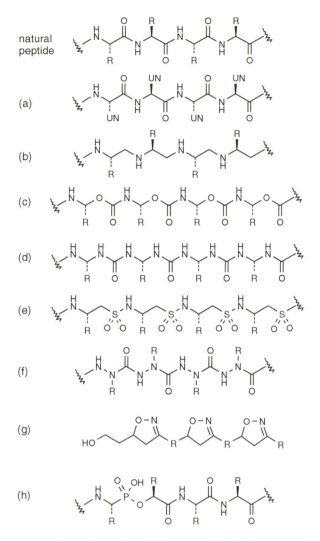

**Fig. 6.6**   A range of oligomeric combinatorial library peptidomimetic products. (UN = unnatural amino acid side-chain.)

syntheses now are routinely labelled as combinatorial approaches. In keeping this book focused on the range and applications of combinatorial chemistry that have been explored, reluctantly this cannot be the place for a comprehensive overview of peptide mimetics.

The following library methods were amongst the first and most influential of the non-peptide combinatorial library approaches (Fig. 6.6). First, there are techniques by which peptides are synthesized from unnatural amino acids (a), and 'library from library' methods for taking the final peptide and chemically converting it to a different oligomer (b). Various groups have replaced the amide bond in natural peptides with carbamates (c), ureas (d), sulfonamides (e), azatides (f), isoxazolines (g), and phosphonates (h). There are many other peptide mimetics but these listed are the ones that so far have been exploited as combinatorial approaches or have the potential to generate combinatorial libraries.

## 4.1   Unnatural peptides

The area of unnatural peptides has been richly exploited for many years as a consequence of the wide availability of unnatural amino acid precursors and the versatility of solid phase peptide synthesis methods. O'Donnell *et al.* (1996) have developed a new approach to these peptides by devising a method for the synthesis of the unnatural amino acids on the solid phase itself, a technique labelled as solid phase unnatural peptide synthesis. To achieve this it was necessary to devise a mild method for the introduction of the side-chains onto the terminal glycine residue on a growing peptide chain. This was accomplished by the alkylation of a Schiff base-activated glycine whilst attached to the solid phase with various electrophiles. A three-step procedure of activation, alkylation, and hydrolysis was added to the normal solid phase peptide synthesis to produce the unnatural peptides (Fig. 6.7) using conditions that were compatible with either the Wang-based or Merrifield-based peptide synthesis methods.

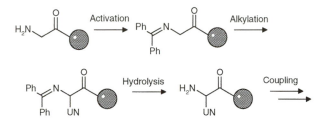

**Fig. 6.7**   The synthesis of unnatural peptides on solid phase depends on the addition of three steps. (UN = unnatural amino acid side-chain.)

The free amine is activated by conversion to the Schiff's base by reaction with benzophenone imine overnight. The activated amino acid could then be alkylated by treatment with a strong non-ionic iminophosphorane base ('Schwesinger base') in the presence of an alkylating agent. Finally, the imine was hydrolysed by aqueous hydrochloric acid/THF, or in the case of Wang resin where the linker is acid-labile, aqueous hydroxylamine hydrochloride at pH 6. The free amine was then available for coupling with another equivalent of Fmoc– or Boc–glycine. This method can also be used for the synthesis of racemic unnatural $\alpha,\alpha$-disubstituted amino acids on solid phase (Scott *et al.* 1997). The procedure outlined above is applied to a solid supported $\alpha$-substituted amino acid to give the disubstituted product following alkylation.

This methodology may be disadvantaged by the number of steps required for the addition of each new amino acid residue. It might have utility if just one position in a peptide chain needed to be diversified with a range of unnatural amino acids, but for longer sequences, it is almost certainly more convenient to prepare the unnatural amino acids before the library synthesis. One further disadvantage is that the alkylation is not stereospecific and all of the new amino acids will have racemic centres. If the synthesis involves two or more synthetic cycles this will lead to complex mixtures of diastereoisomers. O'Donnell however has demonstrated that this approach can be used for the solid phase synthesis of di- and tripeptides and hydantoin libraries on solid phase.

## 4.2 Libraries from libraries

As the structural diversity of peptides might be perceived to be limited, attempts have been made to use peptides as starting points for alternative library structures. Houghten has used an approach he describes as 'libraries from libraries' to convert a library of peptide products into other oligomeric structures by a simple synthetic transformation. For example, a number of pentapeptides on methylbenzhydrylamine (MBHA) solid phase resin were permethylated by reaction with sodium hydride in DMSO followed by addition of methyl iodide (Fig. 6.8) (Ostresh *et al.* 1994). The methylation affected both the backbone amide bonds and the amino acid side-chains in a predictable fashion, with acids converting to the methyl esters, primary amides undergoing methylation to the tertiary amides, and arginine being converted to the trimethylguanidine.

Having demonstrated that the methylation proceeded smoothly, a positional scanning hexapeptide library was subjected to the same conditions and the products tested for antimicrobial activity against two Gram-positive bacteria including *Staphylococcus aureus*.

Houghten has used alkylation conditions to introduce diversity into two positions of a dipeptide library of 57 500 components (Döerner *et al.* 1996). After each coupling step, the solid phase bound compounds were subjected to alkylation of the amide NH to remove any hydrogen bonding capability. As in the permethylation example above, the amino acid side-chains were not always immune to these conditions and also underwent alkylation.

Other groups have explored the possibility of alkylating peptides prepared on solid phase. In a recent paper, a method for the site-selective methylation of peptides has been described (Miller and Scanlan 1997). This method is used in the synthesis of a set of six analogues of the thrombin receptor agonist peptide, $H_2N$–SFLLRN–$CONH_2$ where each of the amide bonds were sequentially and selectively methylated in a process described as '*N*-methyl scanning'. At each stage of the synthesis of the peptide on Rink polystyrene resin, the Fmoc-protecting group was removed and the free terminal amine sulfonylated with *o*-nitrobenzenesulfonyl chloride. The sulfonamide could then be methylated by deprotonation with MTBD and reaction with methyl *p*-nitrobenzenesulfonate. *β*-Mercaptoethanol and DBU removed the sulfonyl group to give the terminal methylamine ready for further amino acid coupling (Fig. 6.9).

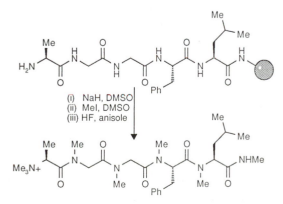

**Fig. 6.8** The trial permethylation of the pentapeptide AGGFL–$NH_2$.

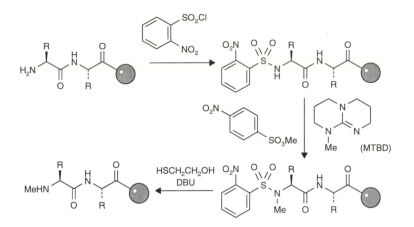

**Fig. 6.9**   The selective methylation of the terminal amine of a solid phase supported peptide.

A further method of producing novel libraries from peptide libraries is through the reduction of the amide bonds by treatment with diborane in THF. Houghten has produced a set of four libraries from libraries by combining reduction and/or alkylation (Houghten *et al.* 1996). After the synthesis of a tetrapeptide library in 52 mixtures of 140 608 peptides, the library was divided into four aliquots each of which was subjected to a different set of conditions. The first was left unchanged, the second was alkylated with excess benzyl bromide to give an *N*-benzylated tetrapeptide library, the third alkylated and reduced with diborane to give an *N*-benzylated pentamine library, and finally the fourth was just reduced to give a pentamine library (Fig. 6.10). These libraries were screened against the (κ-opioid receptor and also in an antimicrobial assay. As might be expected the preferred residues in each position varied depending on the treatment of the peptide libraries, illustrating

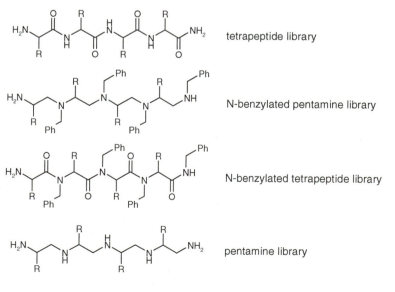

**Fig. 6.10**   The four libraries: the original tetrapeptide library and the three libraries produced from the original library.

in this case that these libraries are not necessarily good peptidomimetics, but instead offer increased structural diversity over a parent peptide library.

## 4.3 Oligocarbamates

Oligocarbamates have been shown to be much more protease-resistant than the native peptide structures they represent whilst at the same time being more lipophilic. The solid phase synthesis of these oligomers from a pool of chiral *N*-protected *p*-nitrophenyl carbonate monomers has been reported (Cho *et al.* 1993). The monomers were prepared from *N*-protected α-amino acids by a borane reduction to the alcohol. A library of these oligomers was prepared using a light-directed parallel synthetic approach (see Chapter 3, Section 3.1) and therefore the light-cleavable NVOC group was chosen for the monomer amine protection. 256 different oligocarbamates were synthesized based on the parent structure AcY$^c$F$^c$A$^c$S$^c$K$^c$I$^c$F$^c$L$^c$ using a binary masking procedure to give a library of all deletion sequences of this oligomer. The nomenclature employed to describe these oligocarbamates uses the one letter amino acid code, indicating that the side-chains reflect the natural peptide sequence, but with a superscript 'c' to indicate a carbamate bond. After each coupling step, half of the glass solid support was irradiated at 365 nm to deprotect the exposed oligomer amine to make it available for coupling in the next synthetic round (Fig. 6.11).

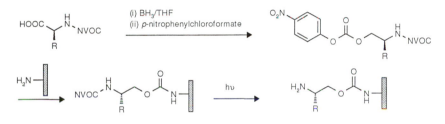

**Fig. 6.11** The light-directed parallel synthesis of oligocarbamates on a glass solid phase support.

The library was screened against a monoclonal antibody, 20D6.3 that had been prepared by immunizing mice with the keyhole limpet haemocyanin conjugate of AcY$^c$K$^c$F$^c$L$^c$G–OH. Using a fluorescent reporter assay, several active sequences were detected in the library and the key recognition epitope was determined to be the sequence–F$^c$L$^c$–. Interestingly, the sequence AcY$^c$F$^c$L$^c$G–OH, although bound by the antibody in solution, was only poorly recognized on the solid support suggesting that the conformation of this ligand may vary depending on whether the compound is on the solid phase or in solution.

In a later paper, Schultz demonstrated that *N*-alkylated oligocarbamates could be readily prepared by the reactions of alternating chiral amino acids and carboxylic acid monomers (Paikoff *et al.* 1996). The synthetic route relies on the acylation of the terminal amino group of the growing oligomer with a carboxylic acid followed by a borane reduction to generate a secondary amine. The carboxylic acid thus becomes an alkyl side-chain on the nitrogen and the amine can be acylated again with an activated *p*-nitrophenylcarbonate of a substituted amino alcohol. The oligocarbamates produced have a repeating chain length of five atoms and two side-chains in each residue (Fig. 6.12).

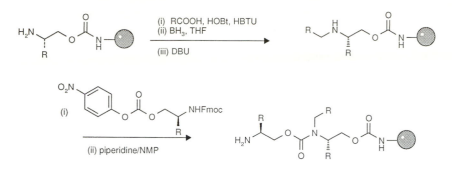

**Fig. 6.12**   *N*-Alkylated oligocarbamates can be synthesized by the iterative cycle of *N*-acylation, reduction, and acylation.

## 4.4   Oligoureas

Ureas are important functional groups in a number of pharmacologically important molecules. As they offer the potential for up to four different substituents, a number of combinatorial approaches to libraries based on a urea template have been explored. This section however describes recent attempts to prepare libraries of peptidomimetics that have a repeating urea element within the backbone replacing the amide bonds in natural peptides. An approach to these molecules has been described by Burgess *et al.* (1995), but as with the oligocarbamate libraries described in the previous section, as the repeating monomer unit contains five atoms, it is difficult to view these oligomers as being deliberate peptide mimics. Rather they are another type of oligomeric structure that has potential to generate biologically active compounds, without any specific attempt to mimic the conformations and spacings of peptide side-chains.

Burgess' approach depended on the availability of monoprotected diamine monomers. These were prepared in solution from Boc-protected α-amino acids by reduction to the amino alcohols followed by Mitsunobu displacement of the hydroxyl with phthalimide and removal of the Boc protecting group. Converting these monomers to the corresponding isocyanates permitted direct addition to the free amine of alanine bound to Rink resin (Fig. 6.13). Deprotection of the phthalimide group proceeded through treatment of the resin with 60% hydrazine hydrate overnight. The free amine-functionalized resin was then available for a further round of monomer addition.

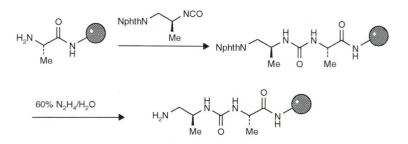

**Fig. 6.13**   The Burgess two-step cycle for the addition of a monomer in the synthesis of oligoureas.

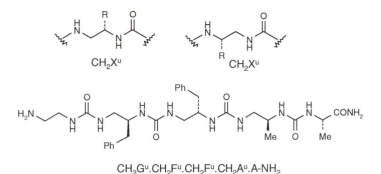

$CH_2G^u.CH_2F^u.CH_2F^u.CH_2A^u.A-NH_2$

**Fig. 6.14** The nomenclature for monomer residues in the oligoureas, and an illustration of its use in describing a prepared compound.

Burgess suggested a shorthand nomenclature for these compounds based on the single letter code for $\alpha$-amino acids. Each residue in the oligomer can be described as '$CH_2X^u$' where 'u' denotes the fact that it is a urea derived fragment, and 'X' refers to the $\alpha$-amino acid used in the monomer synthesis. Depending on the sense of the urea linkage the residue might either be described as '$CH_2X^u$' or '$X^uCH_2$'. Figure 6.14 illustrates this nomenclature with an example that was prepared in an overall yield of 46%.

Burgess *et al.* (1997) have prepared a small combinatorial library of 160 peptidic urea analogues of the Leu enkephalin peptide sequence, $YGGFL-NH_2$ using Rink resin packed into Houghten's tea-bags as the solid support. The library was designed such that each position in the pentapeptide was sequentially replaced with either the D-amino acid, the urea derived from the L-amino acid, or the urea from the D-amino acid. The compounds were cleaved from the Rink resin with a TFA cocktail and screened against a monoclonal antibody that selectively binds the YGGF sequence. It was found that the mAb tolerated urea replacements for the G2 and the L5 positions, but showed no affinity for sequences with other amino acids substituted by ureas. It was also discovered that the shortened tetrameric sequence, $YCH_2G^uFL-NH_2$, had affinity similar to the pentameric compounds, perhaps reflecting the possibility that elongation of the backbone, produced by the introduction of the urea moiety, permitted removal of one amino acid residue.

The interest in oligoureas as pharmacologically promising library compounds is reflected by the work by a second group to make these compounds. Schultz has described a solid phase synthesis of oligoureas on Rink resin that differs in two details (Kim *et al.* 1996*a*). First, rather than activating the monomers as isocyanates, Schultz prepares *para*-nitrophenyl-carbamates. Secondly, the terminal amine introduced by the monomer is not protected as a phthalimide but instead is hidden in the latent form of an azide. Thus the two-step cycle for the introduction of each urea residue requires a tin (II) chloride reduction of the azide prior to the next coupling reaction (Fig. 6.15).

Schultz has also described the iterative synthesis from $\alpha$-amino acids of an oligomeric cyclic urea system (**6.11**) but at present this is a solution phase method which has potential to be transferred to solid phase support (Kim *et al.* 1996*b*).

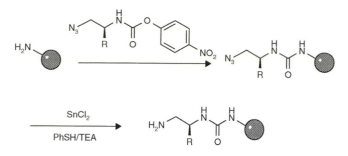

**Fig. 6.15**   The Schultz two-step cycle for the addition of a monomer in the synthesis of oligoureas.

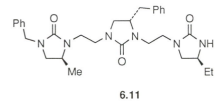

6.11

## 4.5 Peptidosulfonamides

Peptidosulfonamides were initially introduced as transition state mimics of the hydrolysis of the amide bond. The possibility for introducing diversity into these structures through an iterative solid phase synthesis has led to studies by Liskamp into the synthesis of libraries of these oligomers (de Bont *et al.* 1996*a*). This group has prepared peptidosulfonamides on both Merrifield and TentaGel resin, although the latter is claimed to give higher overall chemical yields. For example Fmoc–glycine was loaded onto hydroxy TentaGel resin and the Fmoc group removed with piperidine before coupling with Boc protected aminoethyl sulfinyl chloride (Fig. 6.16). Solid-bound sulfinamide was oxidized by treatment with osmium tetroxide and *N*-methylmorpholine oxide (NMMO) as co-oxidant, although it was found that the yield of the oxidation was highly dependent on the substrate structure.

Liskamp has modified this approach for the synthesis of peptidosulfona-mide-containing peptides based on the structure of Leu-enkephalin (de Bont *et al.* 1996*b*). In order to avoid the problematic oxidation step in the synthesis above, sulfonyl chlorides were used instead of the sulfinyl chlorides (Fig. 6.17). The enkephalin analogues were prepared as carboxylic acids on TentaGel resin using standard solid phase peptide synthesis procedures with

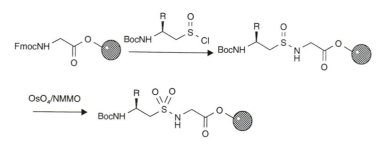

**Fig. 6.16**   The introduction of a peptidosulfonamide monomer on solid phase.

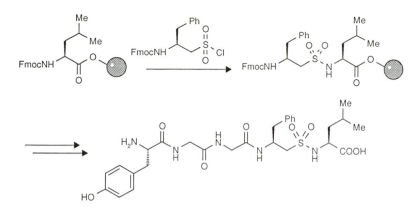

**Fig. 6.17**   The synthesis of a peptidosulfonamide analogue of Leu-enkephalin.

the replacement of an Fmoc–amino acid with the Fmoc-protected aminoethylsulfonyl chloride. A positional scan was used to replace all of the peptide bonds in turn with the sulfonamide bond and following cleavage from the resin, the analogues (terminating in both a carboxylic acid and a carboxamide) were screened against the anti-$\beta$-endorphin monoclonal antibody that recognizes the natural peptide Leu-enkephalin structure. It was found that the sulfonamide replacement was only tolerated at the C-terminus (TyrGlyGlyPheLeu($\Psi CH_2SO_2$)NH$_2$) or between the Phe and Ley residues (TyrGlyGlyPhe($\Psi CH_2SO_2NH$)Leu.NH$_2$).

Underlining the significance of these structures as peptidomimetics and as potential combinatorial library components, a very similar approach to peptidosulfonamides both in solution and on solid phase has been explored by an Italian group (Gude *et al.* 1996). Recognizing the importance of high yields in library synthesis, the coupling cycle developed using 2-aminoalkylsulfonyl chlorides was stated to give a '*combinatorial grade*' conversion of greater than 99%!

## 4.6   Azatides

An oligomeric mimic of peptides has been produced by the replacement of the natural $\alpha$-amino acids with '$\alpha$-aza-amino acids' (Han and Janda 1996). These azatides have been synthesized both in solution and also attached to soluble PEG polymers (see Chapter 4, Section 5). The starting materials employed were a novel alphabet of alkylhydrazines prepared in their Boc-protected forms by the reaction of Boc-carbazate with an aldehyde or ketone followed by hydrogenation of the hydrazone. Alternatively, hydrazine was alkylated prior to Boc protection (Fig 6.18).

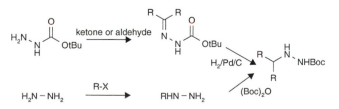

**Fig. 6.18**   The preparation of the key Boc-protected alkylhydrazine monomers required for the synthesis of azatides.

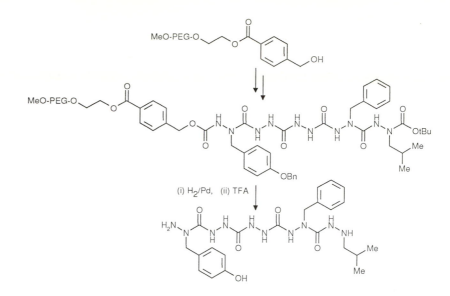

**Fig. 6.19**   The solution polymer-bound synthesis of the azatide mimic of Leu-enkephalin.

Such monomers were used in the synthesis of the azatide mimic of Leu-enkephalin on methoxypolyethylene glycol (Fig. 6.19). The oligomer synthesized (Tyr[a]–Gly[a]–Gly[a]–Phe[a]–Leu[a]) was assessed for its biological activity by screening against the monoclonal antibody 3-E7 that had been raised against the antigen $\beta$-endorphin. Unfortunately the azatide was found not to bind to the mAb, an observation explained by the expectation that the azatide would adopt an extended conformation unlike the $\beta$-turn conformation of the natural peptide.

Despite the inactivity of the enkephalin azatide mimic, the authors hold out the hope that these oligomers may prove useful for the discovery of novel biologically active peptide mimetics.

### 4.7   Polyisoxazolines

Commencing with a polymer-bound olefin it has been proven possible to apply an iterative nitrile oxide 1,3-dipolar cycloaddition and selenide oxidation/elimination to generate polyisoxazolines as peptidomimetics (Kurth *et al.* 1996). Various nitroseleno ethers were dehydrated *in situ* with phenyl isocyanate to give nitrile oxides that were reacted with a terminal olefin attached to carboxylated cross-linked polystyrene. The product isoxazoline attached to the solid phase could then become the substrate for a further cycloaddition following oxidation and elimination of the selenyl group (Fig. 6.20). Using this simple two-step iterative cycle a triisoxazoline was produced in an overall yield of 18%, corresponding with an average yield of 81% per step.

With success in hand, attention was turned to the synthesis of a combinatorial library of 64 triisoxazolines prepared from four readily available nitroseleno ethers and four capping nitro alkanes. A mix and split protocol was used in generating the library compounds as mixtures of four components, and FAB MS was used to demonstrate that the expected ions were present in the final product mixtures.

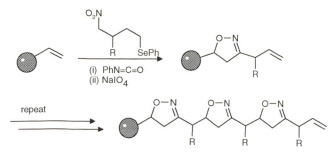

**Fig. 6.20** The application of an iterative two-step procedure for the synthesis of polyisoxazolines.

## 4.8 Peptidyl phosphonates

Peptidyl phosphonates are attractive mimics of peptides with a specific purpose, as the phosphonate group is recognized as an effective transition state mimic for the enzymatic cleavage of the amide bond. Compounds containing this group can be potent inhibitors of peptidases and esterases. A solid phase route to the synthesis of these compounds has been established (Campbell and Bermak 1994). Routine solid phase peptide synthesis was used to attach an Fmoc-protected peptide to resin beads, and following deprotection the amine terminus was acylated with an Fmoc-protected $\alpha$-hydroxy acid (Fig. 6.21). The Fmoc group was removed with piperidine and the hydroxy group converted to a phosphonate by a modified Mitsunobu reaction with a protected aminomethylphosphonate ester. Removal of the amine protecting group with 5% DBU permitted further addition of amino acid residues.

This synthetic approach has now been applied to the preparation of a combinatorial library of peptidyl phosphonates (Campbell *et al.* 1995). The mix and split library procedure was used to prepare a library on amine-substituted TentaGel resin with the intention of screening the compounds against the peptidase, thermolysin, whilst still attached to the solid phase. The 540 library compounds, with the structure Cbz–$X^P$–$^O$Y–Z–resin, were constructed from 18 amino acids (residue Z), five $\alpha$-hydroxy acids (residue Y), and six $\alpha$-aminoalkylphosphonic acids (residue X) using the procedure outlined above.

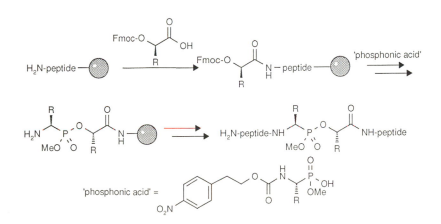

**Fig. 6.21** The synthesis of a peptidyl phosphonate.

**Table 6.1**   The 18 mixtures constructed in the peptoid library[a]

|  |  | Chu = cyclohexylureido |
| --- | --- | --- |
| H–AOD–NH$_2$ | Ac–AOD–NH$_2$ | Chu–AOD–NH$_2$ |
| H–ADO–NH$_2$ | Ac–ADO–NH$_2$ | Chu–ADO–NH$_2$ |
| H–DAO–NH$_2$ | Ac–DAO–NH$_2$ | Chu–DAO–NH$_2$ |
| H–DOA–NH$_2$ | Ac–DOA–NH$_2$ | Chu–DOA–NH$_2$ |
| H–OAD–NH$_2$ | Ac–OAD–NH$_2$ | Chu–OAD–NH$_2$ |
| H–ODA–NH$_2$ | Ac–ODA–NH$_2$ | Chu–ODA–NH$_2$ |

[a] A = aromatic monomers (four variants), O = hydroxylated monomers (three variants), D = diverse monomers (17 variants).

O, A, and D with each of the three capping variants (Table 6.1). As the monomer sets were fixed, each mixture synthesized within the library consisted of 204 trimers (3O × 4A × 17D) plus an additional number of dimers.

The library was synthesized on Rink-derivatized polystyrene resin using the mix and split procedure, with the dimers being generated at the same time by subjecting one mixture in each cycle to a blank reaction. Each final mixture was assayed against a number of 7TM/GPCR targets with routine iterative deconvolution being used to identify the most potent compounds in the library. From this study, several potent ligands were discovered of which CHIR 2279 (**6.12**) was a nanomolar $\alpha_1$–adrenergic receptor ligand, and CHIR 4531 (**6.13**) was a similarly active $\mu$-specific opiate receptor ligand.

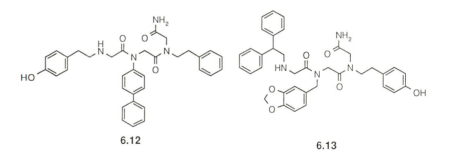

6.12

6.13

## 6.   Summary

The two separate strands in chemistry, that of designing peptide mimetics and the growth of combinatorial chemistry have coincided with the synthesis of libraries of non-natural oligomers. This chapter has described a number of new library structures composed of repeating monomer units, many of which required the development of new solid phase chemistry. The final example of the Chiron *N*-substituted glycines resulted in the discovery of potent biologically active low molecular weight compounds. This is the new direction of combinatorial chemistry—making the tools available that allow the ready synthesis of drug-like molecules in sufficient numbers to irrevocably change the face of drug discovery. One major influence of combinatorial chemistry has been a renaissance in the development of a broad repertoire of organic chemistry on solid phase that allows essentially any molecule to be made through combinatorial synthesis. Chapter 7 describes some key advances from amongst the veritable deluge of new papers describing the solid phase synthesis of organic molecules.

## References

Boons, G-J., Heskamp, B., and Hout, F. (1996). Vinyl glycosides in oligosaccharide synthesis: a strategy for the preparation of trisaccharide libraries based on latent-active glycosylation. *Angew. Chem. Int. Ed. Engl.*, **35**, 2845–7.

Burgess, K., Linthicum, D.S., and Shin, H. (1995). Solid-phase syntheses of unnatural biopolymers containing repeating urea units. *Angew. Chem. Int. Ed. Engl.*, **34**, 907–8.

Burgess, K., Ibarzo, J., Linthicum, D.S., Russell, D.H., Shin, H., Shitangkoon, A., *et al.* (1997). Solid phase syntheses of oligoureas. *J. Am. Chem. Soc.*, **119**, 1556–64.

Campbell, D.A. and Bermak, J.C. (1994). Solid-phase synthesis of peptidylphosphonates. *J. Am. Chem. Soc.*, **116**, 6039–40.

Campbell, D.A., Bernak, J.C., Burkoth, T.S., and Patel, D.V. (1995). A transition state analogue inhibitor combinatorial library. *J. Am. Chem. Soc.*, **117**, 5381–2.

Cho, C.Y., Moran, E.J., Cherry, S.R., Stephens, J.C., Fodor, S.P.A., Adams, C.L., *et al.* (1993). An unnatural biopolymer. *Science*, **261**, 1303–5.

Davis, P.W., Vickers, T.A., Wilson-Lingardo, L., Wyatt, J.R., Guinosso, C.J., Sanghvi, Y.S., *et al.* (1995). Drug leads from combinatorial phosphodiester libraries. *J. Med. Chem.*, **38**, 4363–66.

de Bont, D.B.A., Moree, W.J., and Liskamp, R.M.J. (1996*a*). Molecular diversity of peptidomimetics: approaches to the solid-phase synthesis of peptidosulfonamides. *Bioorg. Med. Chem.*, **4**, 667–72.

de Bont, D.B.A., Dijkstra, G.D.H., den Hartog, J.A.J., and Liskamp, R.M.J. (1996*b*). Solid-phase synthesis of peptidosulfonamide containing peptides derived from leu-enkephalin. *Bioorg. Med. Chem. Lett.*, **6**, 3035–40.

Ding, Y., Labbe, J., Kanie, O., and Hindsgaul, O. (1996). Towards oligosaccharide libraries: a study on the random galactosylation of unprotected *N*-acetylglucosamine. *Bioorg. Med. Chem.*, **4**, 683–92.

Döerner, B., Husar, G.M., Ostresh, J.M., and Houghten, R.A. (1996). The synthesis of peptidomimetic combinatorial libraries through successive amide alkylations. *Bioorg. Med. Chem.*, **4**, 709–15.

Freier, S.M., Konings, D.A.M., Wyatt, J.R., and Ecker, D.J. (1996). 'Mutational SURF': a strategy for improving lead compounds identified from combinatorial libraries. *Bioorg. Med. Chem.*, **4**, 717–25.

Gude, M., Piarulli, U., Potenza, D., Salom, B., and Gennari, C. (1996). A new method for the solution and solid phase synthesis of chiral β–sulfonopeptides under mild conditions. *Tetrahedron Lett.*, **37**, 8589–92.

Han, H. and Janda, K.D. (1996). Azatides: solution and liquid phase syntheses of a new peptidomimetic. *J. Am. Chem. Soc.*, **118**, 2539–44.

Houghten, R.A., Blondelle, S.E., Dooley, C.T., Dörner, B., Eichler, J., and Ostresh, J.M. (1996). Libraries from libraries: generation and comparison of screening profiles. *Mol. Diversity*, **2**, 41–5.

Kanie, O., Barresi, F., Ding, Y., Labbe, J., Otter, A., Forsberg, L.S., *et al.* (1995). A strategy of 'random glycosylation' for the production of oligosaccharide libraries. *Angew. Chem. Int. Ed. Engl.*, **34**, 2720–2.

Kim, J-M., Bi, Y., Paikoff, S.J., and Schultz, P.G. (1996*a*). The solid phase synthesis of oligoureas. *Tetrahedron Lett.*, **37**, 5305–8.

Kim, J-M., Wilson, T.E., Norman, T.C., and Schultz, P.G. (1996*b*). Synthesis of a cyclic urea as a nonnatural biopolymer scaffold. *Tetrahedron Lett.*, **37**, 5309–12.

Kurth, M.J., Randall, L.A.A., and Takenouchi, K. (1996). Solid-phase combinatorial synthesis of polyisoxazolines: a two-reaction iterative protocol. *J. Org. Chem.*, **61**, 8755–61.

Liang, R., Yan, L., Loebach, J., Ge, M., Uozumi, Y., Sekanina, K., *et al.* (1996). Parallel synthesis and screening of a solid phase carbohydrate library. *Science*, **275**, 1520–2.

Miller, S.C. and Scanlan, T.S. (1997). Site–selective *N*-methylation of peptides on solid support. *J. Am. Chem. Soc.*, **119**, 2301–2.

Miller, S.M., Simon, R.J., Ng, S., Zuckermann, R.N., Kerr, J.M., and Moos, W.H. (1994). Proteolytic studies of homologous peptide and *N*-substituted glycine peptoid oligomers. *Bioorg. Med. Chem. Lett.*, **4**, 2657–62.

O'Donnell, M.J., Zhou, C., and Scott, W.L. (1996). Solid-phase unnatural peptide synthesis (UPS). *J. Am. Chem. Soc.*, **118**, 6070–1.

Ostresh, J.M., Husar, G.M., Blondelle, S.E., Dörner, B., Weber, P.A., and Houghten, R.A. (1994). ''Libraries from libraries'': Chemical transformation of combinatorial libraries to extend the range and repertoire of chemical diversity. *Proc. Natl. Acad. Sci. USA*, **91**, 11138–42.

Paikoff, S.J., Wilson, T.E., Cho, C.Y., and Schultz, P.G. (1996). The solid phase synthesis of *N*-alkylcarbamate oligomers. *Tetrahedron Lett.*, **37**, 5653–6.

Schneider, D.J., Feigon, J., Hostomsky, Z., and Gold, L. (1995). High-affinity ssDNA inhibitors of reverse transcriptase of type 1 human immunodeficiency virus. *Biochemistry*, **34**, 9599–610.

Scott, W.L., Zhou, C., Fang, Z., and O'Donnell, M.J. (1997). The solid phase synthesis of $\alpha,\alpha$-disubstituted unnatural amino acids and peptides (di-UPS). *Tetrahedron Lett.*, **38**, 3695–8.

Simon, R.J., Kania, R.S., Zuckermann, R.N., Huebner, V.D., Jewell, D.A., Banville, S., *et al.* (1992). Peptoids: A modular approach to drug discovery. *Proc. Natl. Acad. Sci. USA*, **89**, 9367–71.

Sofia, M.J. (1996). Generation of oligosaccharide and glycoconjugate libraries for drug discovery. *Drug Discovery Today*, **1**, 27–34.

Wyatt, J.R., Vickers, T.A., Roberson, J.L., Buckheit, R.W., Klimkait, T., DeBaets, E., *et al.* (1994). Combinatorially selected guanosine-quartet structure is a potent inhibitor of human immunodeficiency virus envelope-mediated cell fusion. *Proc. Natl. Acad. Sci. USA*, **91**, 1356–60.

Yan, L., Taylor, C.M., Goodnow, R., and Kahne, D. (1994). Glycosylation on the Merrifield resin using anomeric sulfoxides. *J. Am. Chem. Soc.*, **116**, 6953–4.

Zuckermann, R.N., Kerr, J.M., Siani, M.A., Banville, S.C., and Santi, D.V. (1992*a*). Identification of highest-affinity ligands by affinity selection from equimolar peptide mixtures generated by robotic synthesis. *Proc. Natl. Acad. Sci. USA*, **89**, 4505–9.

Zuckermann, R.N., Kerr, J.M., Kent, S.B.H., and Moos, W.H. (1992*b*). Efficient method for the preparation of peptoids [oligo(*N*-substituted glycines)] by submonomer solid-phase synthesis. *J. Am. Chem. Soc.*, **114**, 10646–7.

Zuckermann, R.N., Martin, E.J., Spellmeyer, D.C., Stauber, G.B., Shoemaker, K.R., Kerr, J.M., *et al.* (1994). Discovery of nanomolar ligands for 7-transmembrane G-protein-coupled receptors from a diverse *N*-(substituted)-glycine peptoid library. *J. Med. Chem.*, **37**, 2678–85.

# 7 Solid phase library chemistry

The last chapter documented the ineluctable progression away from the established solid phase synthesis of peptoids, oligonucleotides, and oligo-saccharides towards the novel oligomers demanded by chemists working in combinatorial chemistry. At the end of the chapter, the synthesis of 'peptoids' on solid phase demonstrated that combinatorial chemistry has moved beyond the synthesis of peptide libraries towards the more complicated synthesis of low molecular weight organic molecules. These publications provided an enormous impetus to the further study of solid phase chemistry, and without doubt, in future years they will be considered to have been the watershed that set combinatorial chemistry on its path away from peptide synthesis.

A thorough survey of the solid phase organic synthesis (SPOS) that is now achievable is beyond the scope of this book. Readers who are keen to examine the field in detail are directed to one or more of the excellent reviews of solid phase chemistry (Hermkens *et al.* 1996, 1997). If any measure of the impact of combinatorial chemistry were required, one only has to count the deluge of solid phase synthetic papers. The frequency of publication of new synthetic routes is increasing at an alarming rate; each month a further two dozen new solid phase routes are described. This field is moving so rapidly, any attempt at a comprehensive survey will be frustrated by an exceedingly short shelf-life, and it is the influence of combinatorial chemistry that has forced the pace.

The focus in the first section of this chapter is on the chemistry unique to solid phase: the linker technologies. These varied groups are vital components in any library synthesis, as they will dictate whether the compounds stay on the solid phase for assay, or whether mild and selective conditions are available to cleave the compounds away from the support for solution phase assay. After linkers, there follows a description of the classic solid phase synthetic routes that were responsible for opening up this rich vein of organic synthesis. Finally, I describe some key coupling reactions and convergent approaches to significant molecular types on solid phase. In particular there is a large range of heterocyclic chemistry which appears to be already heavily represented in the literature, and is likely to make a big impact on library design and synthesis in the next five years.

## 1. Linkers

A vital component in any solid phase synthesis strategy is the linker group that tethers the compound being synthesized to the solid phase support. The names of many linkers are irrevocably connected with the type of resin used as support. For example *Merrifield* and *Wang* resins, stalwarts of solid phase peptide synthesis, refer to the specific combination of both the linker and the polymer support itself, although references to the 'Wang linker' are readily understandable.

The earliest linkers were developed for peptide synthesis, and although many of these are still in regular use, the modifications that were made to improve peptide synthesis have also been beneficial for the solid phase synthesis of non-peptides. As chemistry has diverged from the production of the naturally occurring oligomers towards more drug-like molecules, the design, elegance, and subtlety of linkers has advanced in parallel. However, the aims of good linker design remain the same. The linker is a specialized protecting group, in that much of the time, the linker will tie up a functional group, only for it to reappear at the end of the synthesis. The linker needs to be orthoganol. That is, it must not be affected by the chemistry used to modify or extend the attached compound. Finally, the cleavage step should proceed readily and in good yield. As the use of solid phase automatically introduces two new steps into a library synthesis, the best linkers must allow attachment and cleavage in essentially quantitative yields. Furthermore, it is a distinct advantage if the cleavage step can introduce new diversity or is a key step (e.g. cyclization) in the formation of the library products.

Linkers exist that permit the regeneration of the originally linked functionality (alcohols to alcohols), the conversion from one functional group to another (carboxylic acids to amides) or even the total removal of functionality on cleavage (the 'traceless' linkers). It is difficult to bring the vast range of linkers currently used into a systematic listing, as many can be readily applied to the linking of different functionality with minor modification. The categorization below is primarily by the linked functional group, but in some sections it has been convenient to group together the use of one linker (e.g. 2-chlorotrityl linker) for the attachment and generation of several different classes of functional group.

## 1.1 Carboxylic acid linkers

The original linking group used for peptide synthesis bears the name of the father of solid phase peptide synthesis. Merrifield resin is cross-linked polystyrene functionalized with a chloromethyl group (see Chapter 2, Section 2.1) (Merrifield 1963). The attachment of carboxylic acids is usually achieved by the nucleophilic displacement of the chloride with a caesium carboxylate salt in DMF (Fig. 7.1), although any treatise on peptide synthesis will list a variety of methods. Cleavage to regenerate the carboxylic acid is usually achieved with hydrogen fluoride, trifluoromethyl sulfonic acid, or less conveniently by hydrogenolysis. Alcohols can also be linked to Merrifield resin by heating with the corresponding potassium or sodium alkoxide; the addition of a crown ether can be beneficial. Cleavage can be effected by hydrogenolysis or acid treatment in the same manner as in the generation of carboxylic acids.

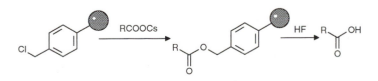

**Fig. 7.1**  Merrifield resin loading and cleavage.

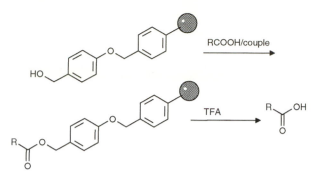

**Fig. 7.2**  The loading and cleavage of the Wang linker.

Merrifield resin is generally unsuitable for the linking of amines or carboxamides as the resulting derivatives cannot easily be cleaved and instead require forcing conditions that could be deleterious to the linked compound. The relative reactivity of Merrifield linked compounds can readily be ascertained by visualizing the products as benzyl derivatives. Whilst benzyl ethers and esters may be cleaved under acidic conditions, benzylamines and *N*-benzylamides are more resistant. As might be assumed, chemists have found a range of alternative linking groups to deal with amines and carboxamides and these are described below.

The second major class of resin linker for carboxylic acids is the Wang linker (Fig. 7.2) (Wang 1973). This linker was designed for the synthesis of peptide carboxylic acids using the Fmoc-protection strategy, and due to the activated benzyl alcohol design—the linker is also know as HMP (hydroxymethylphenoxy resin)—the carboxylic acid products can be readily cleaved with TFA. Although Wang resin refers generally to this linker attached to cross-linked polystyrene, TentaGel and polyacrylamide (as originally developed by Sheppard for peptide synthesis) variants are also commercially available. This resin, often referred to as HMPA (hydroxy-methylphenoxyacetic acid) is linked to aminomethyl TentaGel through an amide bond and thus may have limited applications if strong bases or reducing agents are used.

Wang resin, in addition to being widely used in combinatorial chemistry for the tethering of carboxylic acids is also used as a starting point for other functionality. The ester linkage is vulnerable to nucleophilic attack and thus can be used in the simultaneous cyclization/cleavage approach to diketopiper-azines (Gordon and Steele 1995), benzodiazepines (Mayer *et al.* 1996), and hydantoins (DeWitt *et al.* 1993).

A more acid-labile form of the Wang resin has been developed (Mergler *et al.* 1988). The SASRIN resin (an acronym coined from super acid sensitive resin) has the same structure as the Wang linker but with the addition of a methoxy group to stabilize the carbonium ion formed during acid catalysed cleavage (Fig. 7.3). The resin generally requires only 0.5–1% TFA to generate carboxylic acids and this resin has been used in the solid phase synthesis of biaryl compounds through Suzuki couplings (Guiles *et al.* 1996).

Wang and SASRIN resins have been used for the generation of sulfonamides (Ngu and Patel 1997). Converting the alcohol to a suitable

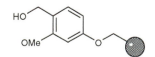

**Fig. 7.3**  The structure of the SASRIN acid-labile linker.

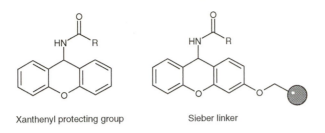

Xanthenyl protecting group          Sieber linker

**Fig. 7.4** The structures of the xanthenyl protecting group for carboxamide amino acid side-chains, and the Sieber carboxamide linker.

leaving group (iodide by Ph$_3$PI$_2$ for example) permitted displacement by an amine. Following sulfonylation with a sulfonyl chloride, TFA catalysed cleavage (95% for Wang polystyrene, 5% for SASRIN linker on TentaGel resin) gave secondary sulfonamides in good yields.

## 1.2   Carboxamide linkers

The Wang ester linker can be cleaved with ammonia to generate primary carboxamides, but this is a difficult reaction that is very slow with sterically hindered amino acids such as valine, and is even further retarded by longer peptide chains. As prolonged treatment with ammonia could lead to racemization of chiral centres, peptide chemists had sought a linking group that would generate carboxamides under mild acidic conditions. The methylbenzhydrylamine linker (MBHA) on polystyrene was first developed for the improved synthesis of peptide amides made on polystyrene using the Boc-protection strategy (Matsueda and Stewart 1981) (Fig. 7.5). At the end of the peptide synthesis the group is cleavable with HF or trifluoromethylsulfonic acid.

With the development of the base-labile Fmoc protection strategy, a carboxamide generating linker with greater acid-lability was keenly sought. One solution to this problem was an acid-labile xanthine derivative developed by Sieber (1987). The aminoxanthenyl group (Fig. 7.4) had been used as acid-labile side-chain protection for the carboxamides in asparagine and glutamine, and Sieber was thus prompted to design a linker based on the same principle. As the xanthenyl group is less acid-labile than Boc, the acid-lability of Sieber's linker was enhanced by the addition of an electron-donating alkoxy substituent, and this same position was used to tether the group to polystyrene solid phase. As a test of the efficiency of this linker, the C-terminal tetrapeptide of human calcitonin, Z–Val–Gly–Ala–Pro.NH$_2$ was prepared in an overall yield of 83%.

An alternative acid-labile carboxamide linker was developed at around the same time. This group, the Rink linker, is now the preferred method for the generation of primary carboxamides on solid phase (Rink 1987) (Fig. 7.5). Rink initially set out to develop an acid-labile linker for carboxylic acids and proposed the hydroxy equivalent linker attached to polystyrene. The paper also describes the conversion of this alcohol to the Rink amide linker by the reaction with 9-fluorenylmethyl carbamate under acidic catalysis. This reaction was also used in the preparation of the Sieber linker, another acid-labile amide linker (see above). The greater acid sensitivity in Rink linkers is a

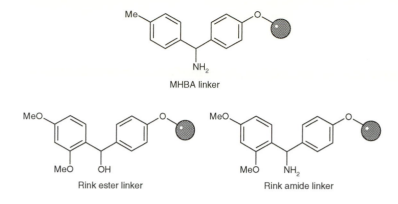

MHBA linker

Rink ester linker          Rink amide linker

**Fig. 7.5**   The structures of the MHBA and Rink linkers.

consequence of the two additional electron-donating methoxy groups. These groups, combined with the ether on the other phenyl ring provided considerable stabilization for the intermediate carbonium ion generated under acidic conditions. In the generation of primary carboxamides, the starting material is attached to the linker as a carboxylic acid and after synthetic modification is cleaved from the resin with TFA.

Rink resin has also been use for the generation of primary sulfonamides (Beaver *et al.* 1996). In an analogous fashion to the generation of primary carboxamides, the Rink amine can be sulfonylated with sulfonyl chlorides. These sulfonamides are resistant to the variety of reaction conditions used in solid phase chemistry, but can be readily cleaved with acids such as 20% TFA in methylene chloride. In a useful modification of this linker, the Rink sulfonamide can be alkylated using potassium *tert*-butoxide and an electrophilic alkyl halide. Acid catalysed cleavage now generates secondary sulfonamides (Fig. 7.6).

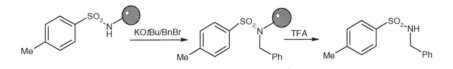

**Fig. 7.6**   Rink resin can be sulfonylated and the intermediate sulfonamides alkylated to give secondary sulfonamides following acid catalysed cleavage.

One of the key objectives of combinatorial chemistry is to find methods for the ready introduction of structural diversity. One option that is frequently exploited is the conversion of a carboxylic acid into a range of carboxamides in the final cleavage step. This is often accomplished by the use of an ester linkage and a final synthetic step that involves reaction with an excess of amine under conditions of elevated temperature. Esters such as the Wang linker however are generally too unreactive to give good yields of amides unless activated. For example, reaction of esters on Wang resin with amines under Lewis acid catalysis has successfully generated amides (Barn *et al.* 1996). A range of catalysts were investigated, but aluminium chloride and zirconium chloride generated the highest yields of amide product.

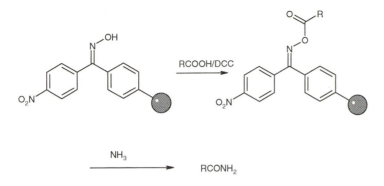

**Fig. 7.7** Kaiser's oxime linker can be coupled with carboxylic acids to generate oxime esters. Aminolysis readily cleaves the ester to generate primary carboxamides.

Other labile esters have been developed for the formation of amides during cleavage. A range of benzophenone oximes were prepared by DeGrado and Kaiser (1980) as amide precursors. The oxime linker allows the formation of oxime esters when coupled to carboxylic acids using dicyclohexylcarbodiimide (DCC) as the coupling reagent. The *p*-nitrobenzophenone oxime ester (Fig. 7.7), generated as a mixture of *syn* and *anti* isomers, is very vulnerable to attack by nucleophiles although essentially stable to 25% TFA and thus is a good candidate for the C-terminal modification of peptides. In particular aminolysis readily generates primary carboxamides from this group, now commonly known as Kaiser's linker. Reaction with hydrazine gives peptide hydrazides, illustrated by the synthesis of BocAsp(OBz)–LeuNHNH$_2$. Finally, reaction of the oxime ester with amino acid esters gave protected peptide esters. Treating oxime resin-bound BocGlu–(OBz) with four equivalents of glycine ethyl ester gave BocGlu(OBz)GlyOEt in 92% chemical yield. Aminolysis of oxime resin is enhanced with catalytic acetic acid, reducing racemization.

Oxime resin has also been used for the introduction of alkylamides onto the C-terminal of peptides (Lobl and Maggiora 1988). Kaiser's oxime resin was used in the synthesis of a section (residues 34–43) of human epidermal growth factor (h-EGF). Treatment with ethylamine or cyclohexylamine in methylene chloride for one to five and a half hours generated the C-terminal amides in high purity. However the oxime linker is labile enough that some chain loss at each step of the peptide synthesis was observed, possibly limiting the use of the resin to shorter synthetic sequences.

Peptides attached to solid phase through Kaiser's oxime resin can also be converted to thioacids by treatment with hexamethyldisilathiane and tetrabutylammonium fluoride (Schwabacher and Maynard 1993). Various oxygen nucleophiles such as alcohols or water will also cleave the oxime linker in the presence of DBU to generate esters or carboxylic acids (Pichette *et al.* 1997). This approach can be used to generate peptide carboxylic acids with the side-chain protection intact; a useful step in the synthesis of more complex peptides, but less useful for combinatorial chemistry.

The above linkers are especially good at generating primary carboxamides. However, in the area of combinatorial library synthesis, it is always advantageous to be able to introduce additional structural diversity during the cleavage step. This has the advantage of both increasing the numbers of compounds in a library, and also may enhance the efficiency of the library method, as the final coupling and cleavage steps have been combined into one. However, as we have seen, the use of amines to cleave a solid phase linker generally requires vigorous reaction conditions. Many amines that the chemist may wish to introduce into the library may lack sufficient nucleophilicity or may be too sterically hindered to react. It is possible of course to enhance the lability of the linker such that it reacts more readily with weakly nucleophilic amines, but the chemist then runs the considerable risk that the linker has only limited stability during the library synthesis.

One solution to this dilemma is the 'safety-catch' linker, a device initially developed for peptide synthesis. The principle is that the initial chemical form of the linker is unreactive to the conditions of the library synthesis but may be activated by a simple chemical transformation to permit cleavage of the library from the solid phase to take place. One such safety-catch is the acylsulfonamide linker developed by Kenner (Kenner *et al.* 1971). Acylsulfonamides are resistant to basic hydrolysis because deprotonation of the acidic NH (p$K_a$ approximately 2.5) generates a stable anion. However, alkylation to the *N*-methyl derivative produces a derivative that can be cleaved under very mild conditions (Fig. 7.8). Kenner extensively chlorosulfonylated cross-linked polystyrene and converted this to the sulfonamide with ammonia. Acylation with a range of Boc-protected amino acids as their 2,4,5-trichlorophenyl esters generated the acylsulfonamides on solid phase. The acylsulfonamide linker is totally stable to the acidic conditions generally used for the removal of Boc or benzyloxycarbonyl groups. However, methylation with diazomethane in ether–acetone produces the activated linker. Alkaline hydrolysis can now generate the carboxylic acid, ammonia give the carboxamide, and hydrazine produce the hydrazide.

**Fig. 7.8**   Kenner's safety-catch is activated by methylation of the acylsulfonamide before nucleophilic cleavage.

The linker has been employed in the solid phase synthesis of non-peptide substituted arylacetic acid derivatives with utility as cyclo-oxygenase inhibitors (Backes and Ellman 1994). Aminomethyl resin was coupled with 4-carboxybenzene sulfonamide and acylation of the sulfonamide with the pentafluorophenyl ester of 4-bromophenylacetic acid produced the 'safe' acylsulfonamide linked substrate (Fig. 7.9). Ellman has demonstrated that the safety-catch is stable to the carbon–carbon bond forming reactions, such as enolate alkylations and palladium catalysed Suzuki couplings, that are becoming increasingly common in solid phase combinatorial chemistry. The final steps in the syntheses were the diazomethane-mediated methylation of the acylsulfonamide followed by cleavage with hydroxide to give carboxylic acids, or amine to generate amides. It was found that benzylamine and

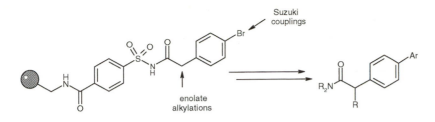

**Fig. 7.9** Phenylacetic acids attached to Kenner's safety-catch linker were alkylated and coupled to boronates before activation of the linker and cleavage with amines.

piperidine gave high yields of the corresponding secondary and tertiary amides, but that aniline would not cleave the linker thus defining the limit of reactivity of the activated safety-catch linker.

Because of the limitations to the reactivity of the methylated safety-catch linker, Ellman has investigated methods of enhancing the reactivity (Backes *et al.* 1996). He hypothesized that the alkylation of the acylsulfonamide with an electron-withdrawing substituent would convert the sulfonamide to a better leaving group and thus accelerate the nucleophilic displacement. The group chosen was cyanomethyl, added through alkylation with bromo- or iodoacetonitrile catalysed with a tertiary amine. The alkylated linker was highly activated towards nucleophilic attack with 0.007 M benzylamine reacting with a half-life of less than five minutes, compared with 790 minutes for the methylated linker. So reactive is this linker that limiting quantities of amine are totally consumed to provide pure amide products. This modified safety-catch linker strategy now provides the opportunity to synthesize combinatorial mixtures of amide products by the addition of equimolar quantities of amines.

### 1.3   Alcohol linkers

A new hydroxyl linker based on the tetrahydropyranyl (THP) protecting group has recently been developed (Thompson and Ellman 1994) (Fig. 7.10). All types of alcohols readily add to dihydropyran and the resulting THP protecting group is stable to strong bases although easily cleaved with acid. The resin is prepared by the reaction of the sodium salt of hydroxymethyldihydropyran with Merrifield resin. Stirring the resin with alcohols under acid catalysis readily generates the linked product. At the end of the synthesis, treatment with 95% TFA in water for 15 minutes regenerates the alcohol.

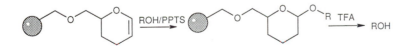

**Fig. 7.10**   The generation and cleavage of the THP linker for alcohols.

The trityl group is well-known as an acid-labile protection group for various heteroatoms. Attachment of the trityl group to a polymer solid phase provides a suitable linker for a range of different functionality including alcohols. Trityl substituted cross-linked polystyrene has been used to anchor

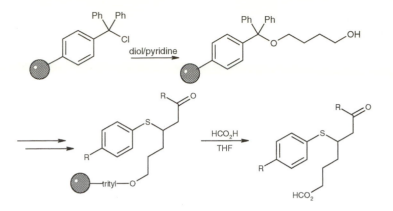

**Fig. 7.11**    The use of trityl resin in the preparation of $\beta$-mercaptoketones.

alcohols in the synthesis of a library of $\beta$-mercaptoketones (Chen *et al.* 1994). The synthesis commenced with the pyridine catalysed addition of a diol to trityl resin, and concluded with formic acid catalysing the cleavage and conversion to the formate ester (Fig. 7.11).

Modified trityl resins have found the considerable utility for linking alcohols and other functionality to the solid phase. Pre-eminent amongst the various forms of trityl linker is 2-chlorotrityl resin. This is commonly used for solid phase synthesis, as the addition of the electron-withdrawing chloride sufficiently reduces lability to allow the formation and ultimate cleavage of alcohols and carboxylic acids (Barlos *et al.* 1989). Reaction of Fmoc–amino acids with chloro 2-chlorotrityl resin proceeds in the presence of diisopropylethylamine (Barlos *et al.* 1991). Cleavage to give the carboxylic acid occurs readily with acetic acid/trifluoroethanol mixtures in a matter of minutes, driven by the enhanced stability of the 2-chlorotrityl cation.

The synthesis of hydroxamates on solid phase has also become available through the 2-chlorotrityl linker. This functionality has become an important goal of combinatorial chemistry as hydroxamates, especially of peptides, can be potent inhibitors of metalloproteinases, and thus have a pharmacological significance. Reaction of 2-chlorotrityl chloride-substituted polystyrene with Fmoc–hydroxylamine and removal of the Fmoc group gives a trityl hydroxylamine that can be readily acylated with Fmoc–amino acids (Mellor *et al.* 1997). Following peptide synthesis, 5% TFA in methylene chloride cleaves the linker to release peptidyl hydroxamic acids (Fig 7.12).

## 1.4    Amine linkers

The first amine linker was used to attach peptides to a solid phase through the N-terminus (Letsinger *et al.* 1964). The dipeptide, LeuGly, was prepared on solid phase linked through the amine, although this has been proven conclusively not to be the preferred direction for peptide synthesis. The carbamate resin used was generated by the reaction of amines with a resin-bound chloroformate prepared from hydroxymethyl polystyrene. Following synthesis, the products were released by the vigorous treatment of the resin with hydrogen bromide in acetic acid. The instability of the chloroformate intermediate has led to the more recent development of activated

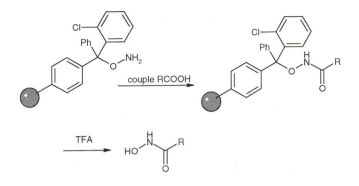

**Fig. 7.12** The synthesis of hydroxamates from 2-chlorotrityl hydroxylamine resin.

*p*-nitrophenylcarbonates as first reported by Leznoff (Dixit and Leznoff 1977). This resin is much more robust and can be stored for at least six months without loss of activity. Reaction with amines is rapid, and after the solid phase synthesis, the amine can be regenerated by treatment with strong acid. The carbamate-linked primary amine can be converted to a secondary methylamine by reaction with lithium aluminium hydride (Ho and Kukla 1997).

Carbamate linkers have been used for the synthesis of a combinatorial library of 576 polyamines prepared in the search of inhibitors of trypanothione reductase, a possible mechanism for the therapy of trypanosomal parasitic infections (Marsh *et al.* 1996). Two linkers were investigated (Fig. 7.13). One based on hydroxymethylbenzoic acid (**7.1**) could be cleaved only with strong acidic conditions (triflic acid/TFA). However, modification of the linker by the addition of a *p*-electron donating group (**7.2**) enhanced the reactivity and allowed cleavage with TFA alone. This latter linkage was preferred for the preparation of libraries destined for solution screening.

Two other primary amine linkers are worthy of note. The carbamate linker has been attached through an allyl spacer, allowing cleavage to be effected by the use of palladium catalysed allyl transfer rather than using acid (Kaljuste and Undén 1996). An alternative non-carbamate linker employs a 4-acetyl-3,5-dioxo-1-methylcyclohexane carboxylic acid (ADCC-linker) (Bannwarth *et al.* 1996). Amines condense readily with the carbonyl of the 4-acetyl group.

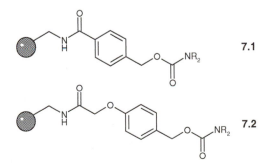

**Fig. 7.13** The two amine bearing linkers used in the preparation of a library of polyamine trypanothione reductase inhibitors.

The linked amine is stable to both acidic and basic conditions but is readily cleaved with hydrazine.

When there are so many options for the attachment of primary amines to solid phase it can be difficult to decide which linker to use. Often the choice is dictated by the needs of the particular library synthesis. For most synthetic uses, the carbamates readily cleaved by acid are most suitable. Clearly, a synthetic route that uses strongly acidic conditions at one point could not use this linker, but might be more attuned to the ADCC-linker.

The carbamate linker is a useful source of a carbonyl group for insertion into heterocycles such as hydantoins (Dressman *et al.* 1996). An amino acid can be coupled through the amine to the carbonate and following conversion of the carboxylic acid to a carboxamide, concomitant cyclization and cleavage can be elicited with triethylamine at elevated temperatures (55–90 °C) (Fig. 7.14). Using this approach a combinatorial library of 800 individual hydantoins have been prepared using 20 amino acids and over 80 primary amines.

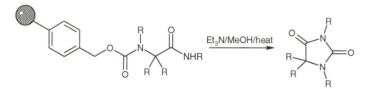

**Fig. 7.14**　The synthesis of hydantoins via a cyclization/cleavage strategy on a carbamate linker.

2-Chlorotrityl resin has been used to link secondary amines and these can be cleaved at the end of the synthesis with 50% TFA (Hoekstra *et al.* 1997). Secondary amines have been attached to Merrifield resin by nucleophilic displacement of the benzyl chloride (Conti *et al.* 1997). However, as noted previously, acid catalysed cleavage from Merrifield is a difficult process, so treatment with α-chloroethyl chloroformate (ACE-Cl) followed by refluxing methanol was successfully used to regenerate the secondary amine. This reaction generates the product as a hydrochloride salt, and carbon dioxide and acetaldehyde dimethyl acetal are also generated as volatile by-products.

A useful linker for the generation of tertiary amines on solid phase has recently been reported (Brown *et al.* 1997). As tertiary amines commonly occur in drug molecules, this method holds significant promise for the synthesis of combinatorial libraries for drug discovery. Primary or secondary amines are introduced to the linker by Michael addition to an acrylate attached to hydroxymethyl polystyrene. Following further synthetic transformations, the amine may be alkylated to give a resin-bound quaternary ammonium ion. Under mildly basic conditions, this system is set-up for a facile Hoffmann elimination to give tertiary amines of high purity (Fig. 7.15). As this approach regenerates the resin-bound acrylate and involves a Michael addition, the support has been labelled as REM resin. The linker requires activation through alkylation before the Hoffmann elimination can take place it could be considered a type of 'safety-catch' linker, although the authors prefer to describe it as a 'traceless' linker (see the next section).

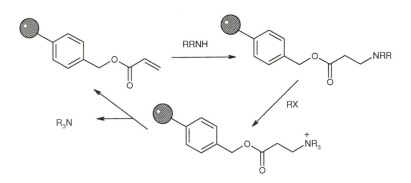

**Fig. 7.15** The generation of tertiary amines by Hoffmann elimination from REM resin beads.

## 1.5 Traceless linkers

The linkers described above depend on the formation of a specific functional group following cleavage from the solid phase. In some cases, the starting materials are loaded onto the resin in one form, such as a carboxylic acid, and cleaved in another form; a carboxamide for example. This is perfectly acceptable if the target compound requires the released functionality. Peptide synthesis lends itself to solid phase chemistry because the molecules invariably contain a carboxylic acid or carboxamide which are obvious anchoring points. However, the growth in interest in combinatorial libraries of low molecular weight non-peptides has elicited a need for new types of linker. Pre-eminent amongst these are linkers that show no specific functionality after cleavage. Such 'traceless' linkers are so-called because an examination of the final compounds reveal no trace of the point of linkage to the solid phase, as it has usually been replaced by an aliphatic or aromatic hydrogen.

The first and by far most widely explored of these traceless linkers is the silyl linker. Silicon attached to a phenyl group can undergo a protodesilylation reaction cleaving the silicon–aryl bond when treated with acid. The benzene ring is readily protonated, as through hyperconjugation, the silicon stabilizes the formation of a $\beta$-carbonium ion. Elimination of silicon releases the final product and restores aromaticity. Several groups in parallel developed the silyl linker based on this principle, but the first to publish were the groups of Veber (Chenera *et al.* 1995) and Ellman (Plunkett and Ellman 1995). Veber's group at SmithKline Beecham linked their silyl linker to Merrifield polystyrene beads and demonstrated palladium catalysed coupling of the aromatic ring before protodesilylation with TFA, caesium fluoride, or liquid HF (Fig. 7.16). Ellman's use of the silyl linker was in the synthesis of a benzodiazepinone on polystyrene beads prior to TFA or HF cleavage.

Several other groups have independently developed an aryl silyl linker, and this is destined to be a major influence in the future design of combinatorial libraries. In addition to protodesilylation, other electrophiles can initiate the cleavage of the aryl–silicon bond. For example, halodesilylation with bromine will replace the silicon with a bromide in the final product. Ellman has also investigated linkers containing the related element, germanium (Plunkett and Ellman 1997). This metal is more readily cleaved than silicon through electrophilic demetallations with halogens to give aryl halides.

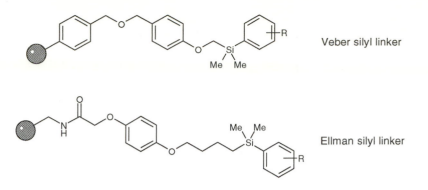

Veber silyl linker

Ellman silyl linker

**Fig. 7.16**  The structures of the traceless silyl linkers used for the generation of aryl derivatives on solid phase.

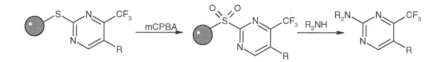

**Fig. 7.17**  The displacement of sulfones from pyrimidines with amine nucleophiles.

Other traceless linkers have been described. A sulfide linker has been used to tether pyrimidines to TentaGel resin (Gayo and Suto 1997). The oxidation of the sulfide to a sulfone with mCPBA allows nucleophilic displacement with amine nucleophiles introducing additional diversity into a library in the final cleavage step (Fig. 7.17). Sulphides have also been used as traceless linkers removed by an oxidation–reduction sequence (Zhao *et al.* 1997). A sulfide starting material has been linked to a soluble PEG polymer, and following synthetic modification, oxidation to the sulfone and reductive cleavage with sodium amalgam generated unsubstituted alkyl compounds in solution.

## 1.6  Light-cleavable linkers

The concept of a light-cleavable linker is highly attractive, as light is only occasionally used in synthetic transformations, and there is consequently less chance of the library chemistry prematurely cleaving products from the solid support. Furthermore, there is an aesthetic appeal to the possibility of irradiating a microtitre plate with UV light at the end of the synthesis to cleave products from resin beads. To this end, successful use has been made of effective photolabile linkers. The first photolabile linker, an *o*-nitrobenzyl resin was developed for the synthesis of individual peptides (Rich and Gurwara 1975). However, this derivative suffers from very slow cleavage rates (12–24 hours), and also generates a chromogenic nitroso aldehyde that may trap liberated products. However, this linker has been used in the synthesis of a variety of tagged amide combinatorial libraries at Pharmacopeia (Baldwin *et al.* 1995). In this case the slow release may have been advantageous as it was possible to release the compound in a controlled manner.

As a consequence of their work in the photolithographic synthesis of peptide libraries (see Chapter 3, Section 3.1), workers at Affymax have

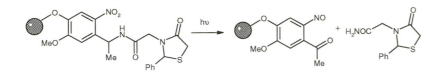

**Fig. 7.18** The use of a photolabile linker to generate a thiazolidinone carboxamide on irradiation.

developed considerable experience of photolabile protecting groups. Following the design of the 6-nitroveratryl protecting groups, they have developed a linker based on α-methyl-6-nitroveratrylamine (Holmes and Jones 1995). This linker, which was shown to be stable to typical TFA deprotection conditions, was used to generate carboxamides from both peptides and small organic molecules on irradiation with UV light. Its versatility was demonstrated in the synthesis of solid support bound 4-thiazolidinones and their release into an aqueous solution suitable for biological assay (Fig. 7.18).

A photolabile nitro-substituted benzhydrylamine linker has been developed for the generation of carboxamides (Ajayaghosh and Pillai 1995).

## 2. The evolution of solid phase chemistry

Solid phase synthesis started with Merrifield's peptide synthesis. Since that time, a whole new field of chemistry has emerged, but the notable feature is the time it has taken for the power of solid phase chemistry to be recognized. The reasons for this are not difficult to fathom. Merrifield established the synthesis of peptides on resin beads, because it was evident to him that the chemistry required to prepare these oligomers was repetitive but reliable. The resin bead method resulted as an enhanced protecting group for the C-terminus, introduced in the first synthetic step and removed almost at the end of the synthesis. In between, were a number of reactions that have now become refined and optimized to the point of formalization. Although there have been attempts to synthesize other oligomers on solid phase, with the possible exception of oligonucleotides, the synthesis of no other family of compounds has been so successfully transferred to the solid phase medium. The measure of the impact of solid phase peptide synthesis is that no alternative synthetic approach would now ever be contemplated.

Between Merrifield's work published in 1963, and the combinatorial explosion in the early 1990s, very few chemists explored the solid phase synthesis of any small organic molecules. Those that did provided a few specific principles to justify their work. Even so, a review from the mid 1970s, had the apologetic title, '*Solid phase organic synthesis: novelty or fundamental concept?*' (Crowley and Rapoport 1976). In this review, the issues surrounding solid phase synthesis were outlined. In particular, the importance of the polymer hyperentropic effect was questioned. It was thought that polystyrene resin maintained sufficient rigidity that reaction sites were isolated from one another giving an effective high dilution effect. However, evidence was mounting that sites did 'talk' to each other and chemistry could be compromised by side-reactions between resin-linked compounds. It is intriguing to note that although this issue was to the fore in 1976, little

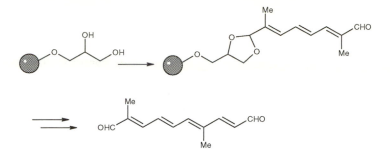

**Fig. 7.19**   The use of solid supports to monoprotect a symmetrical bifunctional molecule in the synthesis of insect sex attractants.

reference to the possibility of by-products from cross-reaction within the resin has been made in the multitude of recent solid phase papers.

Some of the most significant early non-peptide solid phase chemistry was explored by Leznoff (1978). Much use had been made of solid supports to monoprotect symmetrical bifunctional compounds such as diols, dialdehydes, or dicarboxylic acids, a reaction that is comparatively difficult in solution. Leznoff used this approach in the synthesis of various insect sex attractant molecules (Fig. 7.19).

So despite the significance of solid phase peptide synthesis, before 1990, the only other major use of resin beads in synthesis was the monoprotection of difunctional compounds. It took the combined impact of combinatorial peptide library synthesis and the work of Ellman to alert the wider chemical community to the power of solid phase synthesis. Ellman's route to benzodiazepines (Bunin and Ellman 1992) was a highly influential paper. It described an early and successful attempt to prepare a class of recognized drug-like molecules just as the scientific community was latching onto the possibilities of combinatorial chemistry. Whereas in the past most of the power of combinatorial techniques had only been used for the synthesis of peptides, this was a dramatic illustration that the technique could be applied to molecules with proven pharmacological significance. Many realized, possibly for the first time, that a synthesis such as this might be applied to a range of small organic molecules suitable for drug discovery. Such compounds would possess low molecular weights (300–600), have a range of different functionality disposed on a rigid or semi-rigid template, and would avoid metabolically vulnerable groups such as the amide bonds in peptides. Large libraries of such drug-like compounds, prepared through a combinatorial synthetic approach, could offer previously unexploited productivity for the pharmaceutical industry, and generated many new pharmacologically active compounds through high-throughput screening.

Ellman's solid phase route is ideally suited to library design as three discrete monomeric components are combined to form each benzodiazepine (Fig. 7.20). A range of independently synthesized Fmoc-protected 2-aminobenzophenones (**7.3**) were linked to the HMP linker on polystyrene resin through a phenolic or carboxylic acid residue. These derivatives were acylated with a set of Fmoc-protected α-amino acid fluorides, and following deprotection, acid catalysed cyclization gave benzodiazepines (**7.4**). Further

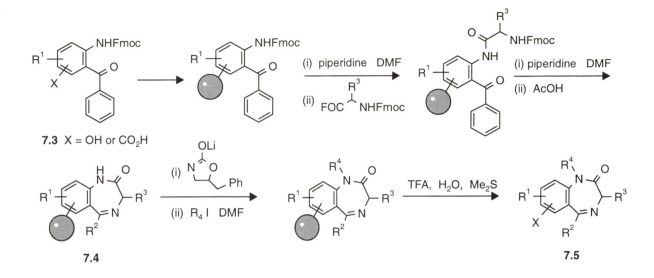

**Fig. 7.20** Ellman's solid phase synthesis of benzodiazepinones.

functionalization was possible by *N*-alkylation of the anilide. The final benzodiazepines (**7.5**) were released from the solid phase by acid treatment with a TFA cocktail and were isolated following chromatography in a relatively pure state ready for screening.

Ellman used this route to prepare a combinatorial library of 192 analogues of the benzodiazepinone structure on multipin solid support (Bunin *et al.* 1994). Clearly Ellman was not alone in spotting the opportunities offered by the solid phase synthesis of libraries of single compounds. Shortly after Ellman's influential paper, a group at Parke-Davis published a similar approach to individual benzodiazepinones using a novel design of apparatus (DeWitt *et al.* 1993).

The Diversomer™ pin apparatus consisted of a number of glass tubes with a porous glass frit at one end, referred to as 'pins' (not to be confused with the grafted polyethylene multipins developed by Geysen and described in Chapter 3, Section 1). These pins were charged with around 100 mg of resin beads for solid phase chemistry. Multiple pins (for example eight in a two by four array) were held by a holder block and the fritted ends were accommodated by a number of separate reaction wells engineered in the 'reservoir' block. Agitation, essential for effective solid phase chemistry, could be achieved by rotational platform shaking, magnetic stirring, or by sonication. The reservoir block could be heated to accelerate chemistry, and a manifold block that fitted over the pins acted as a cooling device for the condensation of heated or refluxing solvents. Reagents were introduced into the top of the pins by automated injection through a gasket-sealed plate. After reaction, the resin samples in each pin were washed by the addition of solvent, repeated agitation, draining under gravity and purging with nitrogen.

Exemplifying their wish to get well away from peptide synthesis, using this apparatus DeWitt *et al.* like Ellman, focused on the preparation of the benzodiazepinone structure, preparing 40 analogues using an original solid

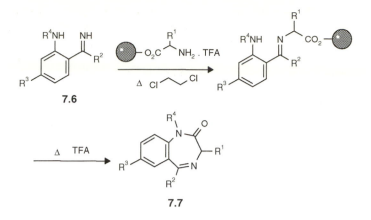

**7.6**

**7.7**

**Fig. 7.21**   The Parke-Davis route to the solid phase synthesis of benzodiazepinones.

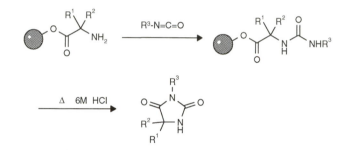

**Fig. 7.22**   The Parke-Davis solid phase synthesis of hydantoins.

phase route. Pre-synthesized aminobenzophenone imines (**7.6**) were trans-iminized using commercially available $\alpha$-amino acids on Wang resin (Fig. 7.21). The resulting imine was cleaved and cyclized by TFA treatment to give (**7.7**).

A similar approach has been devised by the same group for the synthesis of trisubstituted hydantoins (Fig. 7.22), relying on reaction of individual $\alpha$-amino acids with isocyanates followed by an acid hydrolysis. In the quoted example, 39 out of an expected 40 single hydantoins were isolated and characterized.

The Parke-Davis and Ellman groups demonstrated that combinatorial chemistry was not limited to peptides, that it wasn't necessary to make mixtures, and that good yields and purities could be obtained from parallel synthesis. The use of solid phase obviated the opportunity or the need to purify and characterized intermediates, and yet the final products were of acceptable quality for biological screening. No deconvolution was required to identify the products, and the scale of synthesis was sufficient to permit multiple *in vivo* assays.

## 3.   Selected solid phase chemistry

The previous section describes how several key papers in solid phase synthesis alerted the chemical community to the potential scope of combinatorial chemistry. These papers are the classics, primarily as what they reported was previously unexplored, and their publication had a profound effect on the vision and expectations of combinatorial chemists. Since their publication just a few short years ago, the flood gates have opened, and there have been hundreds of papers describing every key type of chemistry being successfully accomplished on resin beads or multipins. The scope of this chemistry is approaching the current range of solution organic chemistry, although naturally there are very few examples, so far, of some types of reaction. Rather than make the futile attempt to offer a complete listing of key solid phase chemistry as applied to combinatorial chemistry, this section aims instead to give a flavour of the areas of synthesis now permissible and in particular, it highlights a few common themes and reactions that are beginning to make a significant impact on library synthesis.

## 3.1 Palladium catalysed couplings

Palladium catalysed chemistry has become highly significant in organic synthesis over the last 20 years. These reactions are popular amongst synthetic chemists because they can be used to form key carbon–carbon bonds and yet proceed under generally mild conditions.

They do not require strong acid or base, or extremes of temperature to work well, and they are now used to achieve reactions, such as aryl–aryl couplings, that are difficult using other methods. Consequently, it comes as no surprise that palladium catalysed reactions are amongst the most widely explored reactions for the synthesis of combinatorial libraries on solid phase.

The formation of aryl–aryl bonds has been established as a key template-forming reaction in combinatorial library synthesis using either the Suzuki or Stille reactions. A halobenzoic acid attached to Merrifield resin was coupled with aryl boronic acids using a range of palladium catalysts (Frenette and Friesen 1994). The reaction is general for a range of aryl halides and aryl boronic acids. A silyl-linked aryl bromide has been reacted with (*p*-formylphenyl)boronic acid under palladium catalysis to give biarylaldehyde products (Chenera *et al.* 1995) (Fig. 7.23). The reaction proceeded to completion in 16 hours and was monitored by Fourier transform infrared spectroscopy (FT–IR) to monitor the appearance of the formyl carbonyl group in the product. The aldehyde could then be used for further derivatization before an acid catalysed protodesilylation freed the products from the resin support.

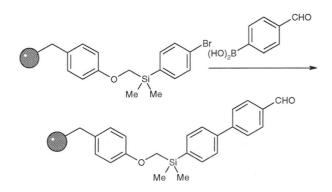

**Fig. 7.23**   The use of the Suzuki aryl–aryl coupling reaction on solid phase.

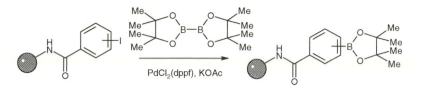

**Fig. 7.24**  Aryl boronates can be generated on solid phase by the reaction of aryl iodides with the pinacol ester of diborane.

Studies have established the generality of the solid phase Suzuki reaction (Guiles *et al.* 1996). In the reaction of a resin-bound aryl bromide with phenyl boronic acid, commercial palladium (0) sources such as tris(dibenzylideneacetone)dipalladium ($Pd_2dba_3$) or tetrakis(triphenylphosphine)palladium (($Ph_3P)_4Pd$) were found to be highly effective, but the choice often depended on the type of boronic acid employed. Alkenyl boronates were coupled more effectively with [1,1'-bis(diphenylphosphino)ferrocene]palladium chloride ($PdCl_2$(dppf)). One limitation of the Suzuki reaction is the limited number of arylboronic acids available for inclusion in a library. To overcome this restriction, resin-bound aryliodides can be converted to boronates by reaction with the pinacol ester of diborane (Piettre and Baltzer 1997) (Fig. 7.24). Suzuki reactions on TentaGel solid phase have been accelerated by the use of microwave irradiation (Larhed *et al.* 1996). The use of microwaves for reactions on solid phase generally has been very poorly explored to date and really deserves a closer and more thorough examination.

The Stille reaction of aryl iodides with alkenyl stannanes is another palladium catalysed reaction that has been developed on solid phase support (Deshpande 1994). 4-Iodobenzoic acid attached to Rink or alanine-derivatized Wang resin was coupled in excellent yields with a range of stannanes using $Pd_2dba_3$ and triphenylarsine (Fig. 7.25).

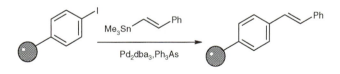

**Fig. 7.25**  The Stille coupling of resin-bound aryl iodides with alkenyl stannanes.

The Stille reaction is also effective at forming aryl–aryl bonds on solid phase (Forman and Sucholeiki 1995). The reaction has been explored in two senses. First, trimethylstannyl phenylacetic acid attached to Rink resin underwent coupling with aryl iodides and aryl triflates using tri-2-furylphosphine (TFP) and $Pd_2dba_3$. Secondly, biaryls were readily formed when the aryl iodide was resin-bound and the arylstannane was in solution.

The Heck reaction is yet another valuable coupling reaction that has successfully been transferred to solid phase. 4-Vinylbenzoic acid or 4-iodobenzoic acid have been attached to Wang resin and coupled with aryl halides/triflates or with alkenes/alkynes respectively under catalysis with $Pd(OAc)_2$ or $Pd_2dba_3$ (Yu *et al.* 1994).

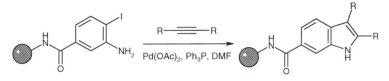

**Fig. 7.26** The palladium catalysed reaction of internal alkynes with *o*-iodoanilines to give diverse indoles.

Palladium chemistry can be used to generate arylamines from aryl bromides and amines (Willoughby and Chapman 1996). The palladium acetate catalysed reaction of alkynes with Rink resin-bound *o*-iodoanilines has been used to synthesize indoles (Zhang *et al.* 1997) (Fig. 7.26).

## 3.2 Mitsunobu couplings

Like the palladium catalysed reactions described above, the Mitsunobu ether-forming reaction is another key coupling reaction that has become a favourite for solid phase chemistry. It allows the formation of aryl ethers under very mild conditions, and is thus unlikely to affect other functionality in the molecule. The first report of this reaction on solid phase was the derivatization of the amino acid tyrosine. A paper from the Selectide Corporation describes the Mitsunobu reaction being used to functionalize *N*-acetylated tyrosine attached to TentaGel resin (Krchnák *et al.* 1995). The chemistry was carried out in a polypropylene syringe using triphenylphosphine and diethyl azodicarboxylate (DEAD). A range of primary and secondary alcohols may be coupled, and this versatility led to the use of this reaction in the preparation of a combinatorial library. The one bead–one compound mix and split strategy was employed to combine 20 natural amino acids, 10 aromatic hydroxy acids, and 21 alcohols in the library of 4200 products (Fig. 7.27).

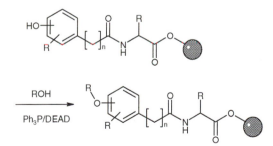

**Fig. 7.27** The generic structure of a 4200 compound library prepared using the Mitsunobu ether coupling reaction.

A group at the Merck laboratories have also reported the Mitsunobu reaction on solid phase. A TentaGel resin-supported benzyl alcohol was coupled with phenols to produce ethers (Rano and Chapman 1995) (Fig. 7.28). The coupling conditions were found to be optimal using a fivefold excess of *N*,*N*,*N*′,*N*′-tetramethylazodicarboxamide (TMAD) and tributylphosphine, and the reaction worked equally well independently of whether the phenol or the benzyl alcohol was attached to the solid support.

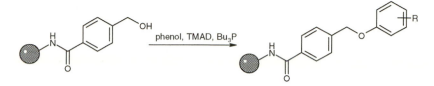

**Fig. 7.28** The formation of ethers on solid phase using the Mitsunobu reaction.

## 3.3 Heterocycle synthesis

Many drug molecules are based on a heterocyclic template. Consequently with the rise in interest in solid phase synthesis of low molecular weight compounds for combinatorial synthesis, many of the approaches have focused on the production of heterocycles. This chapter has already described the classic work on benzodiazepinones by Ellman and on benzodiazepinones and hydantoins by DeWitt. Many other classes of heterocyclic compound have been prepared on solid phase over the last few years.

### 3.3.1 *Benzodiazepines*

Ellman has proceeded to explore the synthesis of other 1,4-benzodiazepines including the 1,4-benzodiazepin-2,5-diones (Boojamra *et al.* 1997). This structure was chosen as it is a 'privileged' structure that occurs in a number of pharmacologically significant molecules. The library was designed to be constructed from several different building blocks, using commercially available monomers where possible, and using synthetic transformations that are compatible with a range of functionality. In fact the library employed 11 alkylating agents, 12 anthranilic acids, and 19 $\alpha$-amino acids (nine sets of enantiomeric pairs and glycine) to give a total of 2508 compounds.

The synthesis was initially attempted on multipins, but more consistent results were obtained by loading the precursors onto Merrifield chloromethyl resin beads, and using a simple 96-well based parallel synthesis apparatus. A resin-bound aldehyde was reductively aminated with an amino acid and then acylated with an anthranilic acid (Fig. 7.29). Cyclization was effected using conditions basic enough to provide an anilide ion for subsequent alkylation. The final benzodiazepinediones were cleaved from the resin with TFA, and the integrity of the compounds ascertained using HPLC on all samples and proton NMR on 36 randomly chosen compounds.

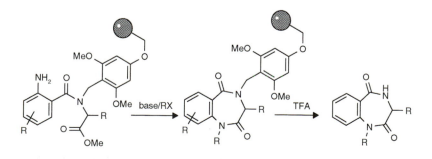

**Fig. 7.29** The solid phase synthesis of benzodiazepinediones.

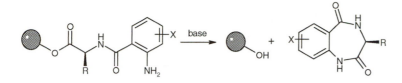

**Fig. 7.30** The cyclization/cleavage route to the synthesis of 1,4-benzodiazepin-2,5-diones.

An earlier paper had reported the synthesis of the same heterocyclic system (Mayer *et al.* 1996). Fmoc–amino acid-derivatized Wang resin had been deprotected and coupled with o-anthranilic acids. The final step in the synthesis was an intramolecular cyclization to form the heterocycle and cleave the molecules from the resin support (Fig. 7.30).

### 3.3.2  Quinolones

The 'Diversomer' technology has been applied by Parke-Davis to the preparation of a small library of eight quinolones (MacDonald *et al.* 1996). Ciprofloxacin is a quinolone antibacterial agent believed to act by the inhibition of DNA gyrase. This compound was initially prepared as a standard for the synthetic approach. Wang resin was derivatized with 2,4,5-trifluorobenzoylacetic acid in the first step of the synthesis. Activation of the $\beta$-ketoester with dimethylformamide dimethyl acetal was followed by addition of cyclopropylamine and formation of the quinolone template was achieved by cyclization under tetramethylguanidine catalysis. There followed another nucleophilic substitution reaction with piperazine to give resin-bound ciprofloxacin (Fig. 7.31). The scope available for the introduction of different amines offers considerable scope for structural variation in the synthesis of a library of these quinolones. The aromatic nucleophilic substitution reaction in particular is becoming another key reaction in solid phase library synthesis (see below).

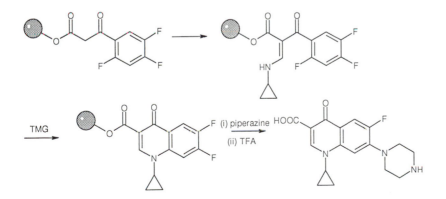

**Fig. 7.31** The solid phase synthesis of ciprofloxacin.

### 3.3.3  The Pictet–Spengler reaction

The Pictet–Spengler reaction is one of the most important intramolecular reactions between iminium ions and nucleophilic aromatic groups. Several

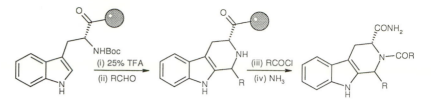

**Fig. 7.32**    The synthesis of β-carbolines on solid phase using the Pictet–Spengler reaction.

groups have described the formation of the β-carboline structure on solid phase from a tryptophan precursor. In one example (Mohan *et al.* 1996), N-Boc tryptophan was linked to Kaiser oxime resin and cyclized with aldehydes under acid catalysis (Fig. 7.32). The product β-carboline could be acylated or sulfonylated to introduce a new centre of diversity before cleavage from the resin with ammonia. In this way ten single analogues were prepared in greater than 71% individual yield.

### 3.3.4    Nucleophilic substitution

A prime objective of combinatorial chemistry is the discovery of methods for the introduction of structural diversity. Therefore to increase the numbers of compounds that can be produced in a library, heterocyclic nucleophilic substitution is an especially attractive type of chemistry. Several groups have investigated the substitution of triazines, pyrimidines and purines with a range of nucleophilic amines. Trichlorotriazine has been coupled to PEG–PS by reaction with a solid-supported amino acid (Fig. 7.33). Because of the decreasing reactivity of tri-, di-, and monochlorotriazines, it was possible to sequentially displace the chlorines with a range of amines to finally generate a library of 12 000 compounds (Stanková and Lebl 1996). 20 resin-bound amino acids were used for the first displacement, and the next two amine displacements were achieved by reacting with 30 and 20 amines respectively at different temperatures.

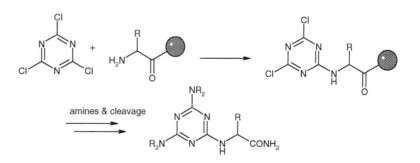

amines & cleavage

**Fig. 7.33**    The successive displacements of chlorine from trichlorotriazine to generate a library of 1200 diverse triazines.

A similar approach has been employed for the synthesis of 2,4,6-trisubstituted pyrimidines (Obrecht *et al.* 1997). One interesting twist in this approach was that the nucleophilic displacement was used in the final step to cleave the pyrimidine product from the solid support (Fig. 7.34). The

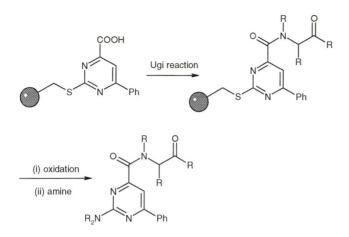

**Fig. 7.34** The synthesis of amine-substituted pyrimidines. The sulfide linker was oxidized to the sulfone prior to the nucleophilic displacement/cleavage step.

pyrimidine was constructed on Merrifield resin such that a sulfide linker connected the heterocycle to the solid phase. At the end of the synthesis, that incidentally used a Ugi reaction to construct $\alpha$-(acylamino)amide (see Section 3.6), the sulfide was oxidized to the sulfone using mCPBA to activate it for nucleophilic displacement. This linker was thus a safety-catch linker, stable to the conditions of the synthesis until it was selectively activated by oxidation.

## 3.4 Cycloadditions

Both Diels–Alder and dipolar cycloadditions have been demonstrated on solid phase. A heterocyclic azadiene attached to carboxylated cross-linked polystyrene has been subjected to cycloaddition with electron-rich dienophiles including both acetylenes and alkenes; a Diels–Alder with reverse electron-demand (Panek and Zhu 1996). Following loss of nitrogen and aromatization, the product 1,2-diazines were generated. These compounds carried a sulfur-based leaving group that could be displaced with amines through a nucleophilic substitution reaction (Fig. 7.35). 16 different derivatives of this structure were produced, ultimately offering the possibility of introducing four sites of diversity.

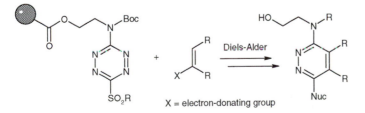

**Fig. 7.35** The synthesis of substituted 1,2-diazines through a reverse electron-demand Diels–Alder reaction.

A Diels–Alder reaction, this time of normal electron-demand, has reacted a benzyl imine on Wang resin with Danishefsky's diene to generate a number of

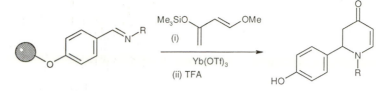

**Fig. 7.36** The solid phase synthesis of 2,3-dihydro-4-pyridones by the Diels–Alder reaction of Danishefsky's diene with imines.

2,3-dihydro-4-pyridones (Wang and Wilson 1997). The cycloaddition was found to proceed in the highest yield when the water tolerant Lewis acid catalyst, Yb(OTf)$_3$ was used (Fig. 7.36). Following acid catalysed cleavage from the resin, the products were isolated in greater than 63% chemical yield and greater than 70% purity.

The solid phase expression of the 1,3-dipolar cycloaddition has been restricted so far to the reactions of nitrile oxides or azomethine ylides (Kantorowski and Kurth 1997). One of the first solid phase 1,3-dipolar cycloadditions to be investigated for library synthesis was used to modify the side-chains of peptide derivatives attached to Rink resin (Pei and Moos 1994). Alkene or alkyne side-chains on peptoid nitrogens were cyclized with reactive nitrile oxides to generate isoxazoles (Fig. 7.37). The nitrile oxides were formed *in situ* from the nitroalkyl compound with phenylisocyanate and triethylamine, or by oxidation of the oximes with sodium hypochlorite.

**Fig. 7.37** The derivatization of peptoid side-chains through 1,3-dipolar cycloaddition reactions.

One of the most effective and striking examples of 1,3-dipolar cycloadditions on solid phase has been the reaction of azomethine ylides with alkenes to give pyrrolidines in a regio- and stereoselective fashion. This reaction has been used to generate a library of 240 proline derivatives using a mix and split protocol on TentaGel resin. The products were tested for inhibitory activity against ACE leading to the identification of subnanomolar inhibitors (Murphy *et al.* 1995). In the first step, an Fmoc-protected α-amino acid (set of four) was coupled to the beads, and after deprotection, an imine was generated from the free amine with one of a set of four aldehydes (Fig. 7.38). This was followed by a silver nitrate catalysed [2+3] cycloaddition with a set of five electron-deficient alkenes. The nitrogen of the proline generated in the cycloaddition was further derivatized with one of three mercaptoacyl groups to complete the synthesis.

The library was screened as mixtures and the most active compound identified by iterative deconvolution. Several active compounds were identified including **7.8**, an ACE inhibitor with a spectacular potency of

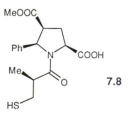

7.8

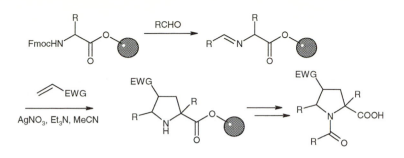

**Fig. 7.38**  The synthesis of an ACE inhibitor library using a 1,3-dipolar cycloaddition as a key step to generate the proline ring.

610 pM. This library has also been prepared using an encoding strategy. The most potent compound was also discovered along with a number of other actives and SAR (see Chapter 5, Section 3.4).

### 3.5   Enolate alkylations

The alkylation of enolate anions is a key reaction in the derivatization of carbon atoms. It plays a pivotal role in solution organic synthesis, and consequently it has been a key reaction to develop on solid phase. However, to date there have been very few examples of enolate reactions achieved on solid phase. One library prepared through this chemistry involved the scandium triflate catalysed reaction of a polymer-supported silyl enol ether with an aldehyde and an amine through a three-component Mannich-type condensation (Kobayashi *et al.* 1996) (Fig. 7.39). 48 individual amino alcohols were prepared through this reaction followed by a lithium borohydride reduction/cleavage.

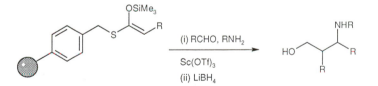

**Fig. 7.39**  The three-component Mannich-type condensation to generate a library of amino alcohols.

### 3.6   The Ugi reaction

Multicomponent reactions, where several different starting materials can be combined in one reaction to give a highly complex product are appealing routes to combinatorial library products. In particular, the Ugi four-component condensation, where a carboxylic acid, an amine, an aldehyde, and an isonitrile generate a substituted $\alpha$-(acylamino)amide (Fig. 7.40), has been the subject of intense investigation for both solution and solid phase library chemistry. It has also been used for the preparation of library compounds ultimately 'captured' onto resin bead supports (see Chapter 4, Section 10).

The size of library that can be generated by the Ugi is limited by the number of available building blocks rather than the number of steps in the

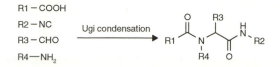

**Fig. 7.40**   The Ugi four-component condensation.

synthetic route. Any of the four components may be attached to solid phase, and the chemistry still proceeds in good yield. For example, the synthesis of Ugi libraries prepared from a solid supported amine have been explored (Tempest *et al.* 1996) (Fig. 7.41). A 96 compound library was prepared as single components from a set of 12 carboxylic acids, eight aldehydes, one isonitrile, and Rink resin as the amine component. The reactions were stirred in methanol and dichloromethane for 24 hours before cleaving the products from the resin with TFA. It was found that 88 of the expected products were obtained in generally good yields.

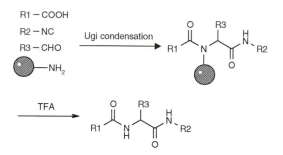

**Fig. 7.41**   The Ugi reaction performed with a solid supported amine component.

As the availability of commercial isonitriles severely limits the diversity of such Ugi-derived libraries, alternative isonitriles were generated in advance of the Ugi condensation by the alkylation of $\alpha$-lithiated benzylisonitrile. A library of 36 components could then be synthesized from three carboxylic acids, three aldehydes, and four isonitriles. An alternative solution to the isonitrile availability problem is the use of 1-isocyanocyclohexene (**7.9**) which can be used as a precursor to a range of other functionality following Ugi reactions (Strocker *et al.* 1996).

The Ugi condensation has been employed by the same research group to prepare a combinatorial library of sialyl Lewis x glycomimetics (Sutherlin *et al.* 1996). Using one fucose derivative as the aldehyde component, a set of six amines on Rink resin (five $\alpha$-amino acids and unsubstituted Rink itself), two isonitriles, and eight symmetrical dicarboxylic acids, 192 individual Ugi products were generated (Fig. 7.42). The final products were treated with sodium hydroxide to cleave the fucose protecting acetyl groups and cleaved from the Rink resin with TFA. Nearly all of the expected products were formed in approximately 50% yield and in high purity, demonstrating the power and simplicity of multicomponent condensations on solid phase.

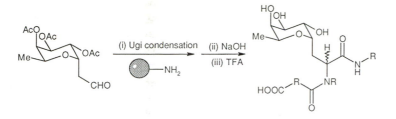

**Fig. 7.42** The synthesis of a library of sialyl Lewis x mimetics through the Ugi reaction on solid phase.

## 4. Summary

This chapter has given little more than a flavour of the range of chemistry that has now been transferred to solid phase, and in particular, highlights those chemistries that have already been used for the preparation of combinatorial libraries. There is no doubt that this is currently the major growth area in organic synthesis. As more advances are made in preparing compounds on solid phase this area of chemistry that was neglected for so long will make contributions to the synthesis of organic molecules both within and beyond combinatorial synthesis.

## References

Ajayaghosh, A. and Pillai, V.N.R. (1995). Solid-phase synthesis and C-terminal amidation of peptides using a photolabile *o*-nitrobenzhydrylaminopolystyrene support. *Tetrahedron Lett.*, **36**, 777–80.

Backes, B.J. and Ellman, J.A. (1994). Carbon–carbon bond-forming methods on solid support. Utilization of Kenner's 'safety-catch' linker. *J. Am. Chem. Soc.*, **116**, 11171–2.

Backes, B.J., Virgilio, A.A., and Ellman, J.A. (1996). Activation method to prepare a highly reactive acylsulphonamide 'safety-catch' linker for solid-phase synthesis. *J. Am. Chem. Soc.*, **118**, 3055–6.

Baldwin, J.J., Burbaum, J.J., Henderson, I., and Ohlmeyer, M.H.J. (1995). Synthesis of a small molecule combinatorial library encoded with molecular tags. *J. Am. Chem. Soc.*, **117**, 5588–9.

Bannwarth, W., Huebscher, J., and Barner, R. (1996). A new linker for primary amines applicable to combinatorial approaches. *Bioorg. Med. Chem. Lett.*, **6**, 1525–8.

Barlos, K., Gatos, D., Kapolos, S., Papaphotiu, G., Schäfer, W., and Wenqing, Y. (1989). Veresterung von partiell geschützten Peptid-fragmenten mit Harzen. Einsatz von 2-Chlortritylchlorid zur Synthese von leu15-Gastrin I. *Tetrahedron Lett.*, **30**, 3947–50.

Barlos, K., Chatzi, O., Gatos, D., and Stavropoulos, G. (1991). 2-Chlorotrityl chloride resin. *Int. J. Pept. Protein Res.*, **37**, 513–20.

Barn, R., Morphy, J.R., and Rees, D.C. (1996). Synthesis of an array of amides by aluminium chloride assisted cleavage of resin-bound esters. *Tetrahedron Lett.*, **37**, 3213–16.

Beaver, K.A., Siemund, A.C., and Spear, K.L. (1996). Application of the sulfonamide functional group as an anchor for solid phase organic synthesis (SPOS). *Tetrahedron Lett.*, **37**, 1145–8.

Boojamra, C.G., Burow, K.M., Thompson, L.A., and Ellman, J.A. (1997). Solid-phase synthesis of 1,4-benzodiazepine-2,5-diones. Library preparation and demonstration of synthesis generality. *J. Org. Chem.,* **62**, 1240–56.

Brown, A.R., Rees, D.C., Rankovic, Z., and Morphy, J.R. (1997). Synthesis of tertiary amines using a polystyrene (REM) resin. *J. Am. Chem. Soc.*, **119**, 3288–95.

Bunin, B.A. and Ellman, J.A. (1992). A general and expedient method for the solid-phase synthesis of 1,4-benzodiazepine derivatives. *J. Am. Chem. Soc.*, **114**, 10997–8.

Bunin, B.A., Plunkett, M.J., and Ellman, J.A. (1994). The combinatorial synthesis and chemical and biological evaluation of a 1,4-benzodiazepine library. *Proc. Natl. Acad. Sci. USA*, **91**, 4708–12.

Chen, C., Randall, L.A.A., Miller, R.B., Jones, A.D., and Kurth, M.J. (1994). 'Analogous' organic synthesis of small-compound libraries: Validation of combinatorial chemistry in small-molecule synthesis. *J. Am. Chem. Soc.*, **116**, 2661–2.

Chenera, B., Finkelstein, J.A., and Veber, D.F. (1995). Protodetachable arylsilane polymer linkages for use in solid-phase organic synthesis. *J. Am. Chem. Soc.*, **117**, 11999–2000.

Conti, P., Demont, D., Cals, J., Ottenheijm, H.C.J., and Leysen, D. (1997). A new cleavage strategy for the solid-phase synthesis of secondary amines. *Tetrahedron Lett.*, **38**, 2915–18.

Crowley, J.I. and Rapoport, H. (1976). Solid-phase organic synthesis: novelty or fundamental concept? *Acc. Chem. Res.*, **9**, 135–44.

DeGrado, W.F. and Kaiser, E.T. (1980). Polymer-bound oxime esters as supports for solid-phase peptide synthesis. Preparation of protected peptide fragments. *J. Org. Chem.*, **45**, 1295–300.

Deshpande, M.S. (1994). Formation of carbon–carbon bond on solid support: application of the Stille reaction. *Tetrahedron Lett.*, **35**, 5613–14.

DeWitt, S.H., Kiely, J.S., Stankovic, C.J., Schroeder, M.C., Cody, D.M.R., *et al.* (1993). 'Diversomers': An approach to nonpeptide, nonoligomeric chemical diversity. *Proc. Natl. Acad. Sci. USA*, **90**, 6909–13.

Dixit, D.M. and Leznoff, C.C. (1977). Insoluble polymer supports as monoblocking agents of symmetrical diamines. *J. Chem. Soc. Chem. Commun.*

Dressman, B.A., Spangle, L.A., and Kaldor, S.W. (1996). Solid phase synthesis of hydantoins using a carbamate linker and a novel cyclization/cleavage step. *Tetrahedron Lett.*, **37**, 937–40.

Forman, F.W. and Sucholeiki, I. (1995). Solid-phase synthesis of biaryls via the Stille reaction. *J. Org. Chem.*, **60**, 523–8.

Frenette, R. and Friesen, R.W. (1994). Biaryl synthesis via Suzuki coupling on a solid support. *Tetrahedron Lett.*, **35,** 9177–80.

Gayo, L.M. and Suto, M.J. (1997). Traceless linker: oxidative activation and displacement of a sulfur-based linker. *Tetrahedron Lett.*, **38**, 211–14.

Gordon, D.W. and Steele, J. (1995). Reductive alkylation on a solid phase: synthesis of a piperazinedione combinatorial library. *Bioorg. Med Chem. Lett.*, **5**, 47–50.

Guiles, J.W., Johnson, S.G., and Murray, W.V. (1996). Solid-phase Suzuki coupling for C-C bond formation. *J. Org. Chem.*, **61**, 5169–71.

Hermkens, P.H.H., Ottenheijm, H.C.J., and Rees, D. (1996). Solid-phase organic reactions: A review of the recent literature. *Tetrahedron*, **52**, 4527–54.

Hermkens, P.H.H., Ottenheijm, H.C.J., and Rees, D.C. (1997). Solid-phase organic reaction II. A review of the literature Nov 95–Nov 96. *Tetrahedron*, **53**, 5643–78.

Ho, C.Y. and Kukla, M.J. (1997). Carbamate linkers as latent *N*-methylamines in solid phase synthesis. *Tetrahedron Lett.*, **38**, 2799–802.

Hoekstra, W.J., Greco, M.N., Yabut, S.C., Hulshizer, B.L., and Maryanoff, B.E. (1997). Solid-phase synthesis via N-terminal attachment to the 2-chlorotrityl resin. *Tetrahedron Lett.*, **38**, 2629–32.

Holmes, C.P. and Jones, D.G. (1995). Reagents for combinatorial organic synthesis: development of a new *o*-nitrobenzyl photolabile linker for solid phase synthesis. *J. Org. Chem.*, **60**, 2318–19.

Kaljuste, K. and Undén, A. (1996). Solid-phase synthesis of peptide aminoalkylamides using an allyl linker. *Tetrahedron Lett.*, **37**, 3031–4.

Kantorowski, E.J. and Kurth, M.J. (1997). Dipolar cycloadditions in solid-phase organic synthesis (SPOS). *Mol. Diversity*, **2**, 207–16.

Kenner, G.W., McDermott, J.R., and Sheppard, R.C. (1971). The safety catch principle in solid-phase synthesis. *J. Chem. Soc. Chem. Commun.*

Kobayashi, S., Moriwaki, M., Akiyama, R., Suzuki, S., and Hachiya, I. (1996). Parallel synthesis using Mannich-type three-component reactions and 'Field synthesis' for the construction of an amino alcohol library. *Tetrahedron Lett.*, **37**, 7783–6.

Krchnák, V., Flegelová, Z., Weichsel, A.S., and Lebl, M. (1995). Polymer-supported Mitsunobu ether formation and its use in combinatorial chemistry. *Tetrahedron Lett.*, **36**, 6193–6.

Larhed, M., Lindeberg, G., and Hallberg, A. (1996). Rapid microwave-assisted Suzuki coupling on solid-phase. *Tetrahedron Lett.*, **37**, 8219–22.

Letsinger, R.L., Kornet, M.J., Mahadevan, V., and Jerina, D.M. (1964). Reactions on polymer supports. *J. Am. Chem. Soc.*, **86**, 5163–5.

Leznoff, C.C. (1978). The use of insoluble polymer supports in general organic synthesis. *Acc. Chem. Res.*, **11**, 327–33.

Lobl, T.J. and Maggiora, L.L. (1988). Convenient synthesis of C-terminal peptide analogues by aminolysis of oxime resin-linked protected peptides. *J. Org. Chem.*, **53**, 1979–82.

MacDonald, A.A., DeWitt, S.H., Hogan, E.M., and Ramage, R. (1996). A solid phase approach to quinolones using the Diversomer Technology. *Tetrahedron Lett.*, **37**, 4815–18.

Marsh, I.R., Smith, H., and Bradley, M. (1996). Solid phase polyamine linkers—their utility in synthesis and the preparation of directed libraries against trypanothione reductase. *J. Chem. Soc. Chem. Commun.*

Matsueda, G.R. and Stewart, J.M. (1981). A *p*-methylbenzhydrylamine resin for improved solid-phase synthesis of peptide amides. *Peptides*, **2**, 45–50.

Mayer, J.P., Zhang, J., Bjergarde, K., Lenz, D.M., and Gaudino, J.J. (1996). Solid phase synthesis of 1,4-benzodiazepine-2,5-diones. *Tetrahedron Lett.*, **37**, 8081–4.

Mellor, S.L., McGuire, C., and Chan, W.C. (1997). *N*-Fmoc-aminooxy-2-chlorotrityl polystyrene resin: A facile solid-phase methodology for the synthesis of hydroxamic acids. *Tetrahedron Lett.*, **38**, 3311–14.

Mergler, M., Tanner, R., Gosteli, J., and Grogg, P. (1988). Peptide synthesis by a combination of solid-phase and solution methods I: a new very acid-labile anchor group for the solid-phase synthesis of fully protected fragments. *Tetrahedron Lett.*, **29**, 4005–8.

Merrifield, R.B. (1963). Solid phase peptide synthesis. I. The synthesis of a tetrapeptide. *J. Am. Chem. Soc.*, **85**, 2149–54.

Mohan, R., Chou, Y-L., and Morrissey, M.M. (1996). Pictet-Spengler reaction on solid support: synthesis of 1,2,3,4-tetrahydro-$\beta$-carboline libraries. *Tetrahedron Lett.*, **37**, 3963–6.

Murphy, M.M., Schullek, J.R., Gordon, E.M., and Gallop, M.A. (1995). Combinatorial organic synthesis of highly functionalized pyrrolidines: identification of a potent angiotensin converting enzyme inhibitor from a mercaptoacyl proline library. *J. Am. Chem. Soc.*, **117**, 7029–30.

Ngu, K. and Patel, D.V. (1997). Preparation of acid-labile resins with halide linkers and their utility in solid-phase organic synthesis. *Tetrahedron Lett.*, **38**, 973–6.

Obrecht, D., Abrecht, C., Grieder, A., and Villalgordo, J.M. (1997). A novel and efficient approach for the combinatorial synthesis of structurally diverse pyrimidines on solid support. *Helv. Chim. Acta*, **80**, 65–72.

Panek, J.S. and Zhu, B. (1996). Synthesis of aromatic 1,2-diazines by inverse electron demand Diels-Alder reaction of polymer-supported 1,2,4,5-tetrazines. *Tetrahedron Lett.*, **37**, 8151–4.

Pei, Y. and Moos, W.H. (1994). Post-modification of peptoid side chains: [3+2] cycloaddition of nitrile oxides with alkenes and alkynes on the solid-phase. *Tetrahedron Lett.*, **35**, 5825–8.

Pichette, A., Voyer, N., Larouche, R., and Meillon, J-C. (1997). A useful method for the preparation of fully protected peptide acids and esters. *Tetrahedron Lett.*, **38**, 1279–82.

Piettre, S.R. and Baltzer, S. (1997). A new approach to the solid-phase Suzuki coupling reaction. *Tetrahedron Lett.*, **38**, 1197–200.

Plunkett, M.J. and Ellman, J.A. (1995). A silicon-based linker for traceless solid-phase synthesis. *J. Org. Chem.*, **60**, 6006–7.

Plunkett, M.J. and Ellman, J.A. (1997). Germanium and silicon linking strategies for traceless solid-phase synthesis. *J. Org. Chem.*, **62**, 2885–93.

Rano, T.A. and Chapman, K.T. (1995). Solid phase synthesis of aryl ethers via the Mitsunobu reaction. *Tetrahedron Lett.*, **36**, 3789–92.

Rich, D.H. and Gurwara, S.K. (1975). Preparation of a new *o*-nitrobenzyl resin for solid-phase synthesis of *tert*-butyloxycarbonyl-protected peptide acids. *J. Am. Chem. Soc.*, **97**, 1575–9.

Rink, H. (1987). Solid-phase synthesis of protected peptide fragments using a trialkoxy-diphenyl-methylester resin. *Tetrahedron Lett.*, **28**, 3787–90.

Schwabacher, A.W. and Maynard, T.L. (1993). Preparation of peptide thioacids using Kaiser oxime resin. *Tetrahedron Lett.*, **34**, 1269–70.

Sieber, P. (1987). A new acid-labile anchor group for the solid-phase synthesis of C-terminal peptide amides by the Fmoc method. *Tetrahedron Lett.*, **28**, 2107–10.

Stanková, M. and Lebl, M. (1996). Library generation through successive substitution of trichlorotriazine. *Mol. Diversity*, **2**, 75–80.

Strocker, A.M., Keating, T.A., Tempest, P.A., and Armstrong, R.W. (1996). Use of a convertible isocyanate for generation of Ugi reaction derivatives on solid support: synthesis of α-acylaminoesters and pyrroles. *Tetrahedron Lett.*, **37**, 1149–52.

Sutherlin, D.P., Stark, T.M., Hughes, R., and Armstrong, R.W. (1996). Generation of C-glycoside peptide ligands for cell surface carbohydrate receptors using a four-component condensation on solid support. *J. Org. Chem.*, **61**, 8350–4.

Tempest, P.A., Brown, S.D., and Armstrong, R.W. (1996). Solid-phase parallel syntheses by Ugi multicomponent condensation. *Angew. Chem. Int. Ed. Engl.*, **35**, 640–2.

Thompson, L.A. and Ellman, J.A. (1994). Straightforward and general method for coupling alcohols to solid supports. *Tetrahedron Lett.*, **35**, 9333–6.

Wang, S-S. (1973). *p*-Alkoxy alcohol resin and *p*-alkoxybenzyloxycarbonyl-hydrazine resin for solid phase synthesis of protected peptide fragments. *J. Am. Chem. Soc.*, **95**, 1328.

Wang, Y. and Wilson, S.R. (1997). Solid phase synthesis of 2,3-dihydro-4-pyridones: reaction of Danishefsky's diene with polymer-bound imines. *Tetrahedron Lett.*, **38**, 4021–4.

Willoughby, C.A. and Chapman, K.T. (1996). Solid phase synthesis of aryl amines. *Tetrahedron Lett.*, **37**, 7181–4.

Yu, K-L., Deshpande, M.S., and Vyas, D.M. (1994). Heck reactions in solid phase synthesis. *Tetrahedron Lett.*, **35**, 8919–22.

Zhang, H-C., Brumfield, K.K., and Maryanoff, B.E. (1997). Synthesis of trisubstituted indoles on the solid-phase via palladium-mediated heteroannulation of internal alkynes. *Tetrahedron Lett.*, **38**, 2439–42.

Zhao, X-Y., Jung, K.W., and Janda, K.D. (1997). Soluble polymer synthesis: an improved traceless linker methodology for aliphatic C–H bond formation. *Tetrahedron Lett.*, **38**, 977–80.

# 8 Analysis of chemistry and products

Many aspects of combinatorial chemistry are merely a more specialized extension of existing chemical technology. We have seen in previous chapters that combinatorial chemistry emerged as a reaction to the needs of immunochemists to prepare more peptides for epitope studies, and that this technique has been widely adopted by scientists in other disciplines who have seen the obvious benefits of high synthetic productivity. Rapid synthesis has been used to accelerate the study of the solid phase synthesis of peptidomimetics and oligosaccharides, and more recently, as described in the previous chapter, the range of accessible solid phase chemistry of small molecules has been enlarged as a response to the possibilities of combinatorial chemistry.

Exploring new avenues of solid phase chemistry presents new challenges, not least of which is the question of how to monitor chemistry as it takes place, and how to confirm that a compound has been synthesized as planned. In parallel with the colossal growth in solid phase synthesis of low molecular weight compounds, there has also been a renaissance in the techniques for analysis of compounds whilst still attached to the solid phase. These techniques are not new, rather combinatorial chemistry has spurred a more focused study of chromatographic and spectroscopic techniques that meet the requirements of synthesis on resin beads and other solid supports.

This topic can be divided into three main areas. First, there are those techniques that monitor chemical conversion taking place on the resin bead, primarily as a guide to when chemical conversion is complete and a reaction can be worked-up. This is especially useful as the techniques beloved of the solution chemist, thin-layer chromatography especially, are no longer available if the product is attached to the solid phase. Secondly, there are the techniques that allow the analysis of the reaction products, again avoiding if possible the chemical cleavage from the solid phase. Lastly, there are those techniques that permit an analysis of a combinatorial mixture of compounds. Although this last objective might seem to be exceptionally challenging, new techniques allow confirmation of the presence of large proportions of the compounds in high order mixtures. There are even methods now available for the simultaneous screening and analysis of mixtures.

## 1. Analysing synthetic success

Monitoring the completeness of solid phase reactions is straightforward when the main concern of the combinatorial chemist is the total consumption of available amino groups. The Kaiser test continues to be the most widely known method for assessing the presence of primary amines on solid support.

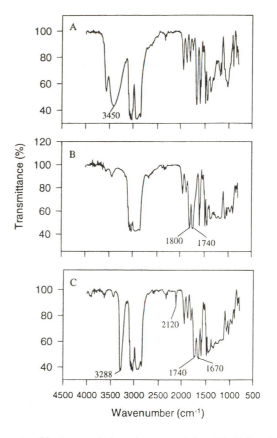

**Fig. 8.2**   IR spectra of the three resin-bound compounds from Fig. 8.1. Reprinted with permission from *J. Org. Chem.* (1995), **60**, 5736–8. Copyright 1995 American Chemical Society.

and esterification (Yan *et al.* 1996*a*), and the tetra-*n*-propylammonium perruthenate (TPAP) oxidation of primary and secondary alcohols (Yan *et al.* 1996*b*). The study of the $S_N2$ reaction was carried out by removing beads from the reaction at various time points, washing the beads, and observing the IR spectrum of single flattened beads. In this way it was possible to ascertain the pseudo-first order rate constant for the displacement of a resin-bound benzyl chloride with potassium acetate. It was found that the rate of this reaction on solid phase is much faster than the rate of similar solution phase reactions, possibly as a consequence of the high local concentrations of reactant in the resin bead. In a further experiment it was shown that reactions on TentaGel resin proceed no faster than the same reactions on Wang resin beads.

The paper also describes the use of attenuated total reflection (ATR) as a technique to examine the reaction rate of functionality on the surface of the bead. More than 99% of the total chemical functionality is sited inside the bead, and it is presumed that chemistry here will be subject to diffusion control. Transmission FT–IR microspectroscopy of the flattened bead indicates the progression of chemistry inside the bead, whereas placing the ATR objective in physical contact with the bead surface records a spectrum of material that is mainly on the surface.

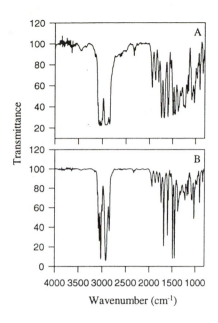

**Fig. 8.3**   IR spectra from a single Merrifield resin bead (A) and a flattened bead (B). Reprinted from *Tetrahedron*, **52**, B. Yan and G. Kumaravel, Probing solid phase reactions by monitoring the IR bands of compounds on a single 'flattened' bead, 843–8, Copyright 1996, with kind permission from Elsevier Science Ltd, The Boulevard, Langford Lane, Kidlington, OX5 1GB, UK.

A comparison was made of the rates of esterification of Wang resin with phenylacetic acid/DIC/DMAP both on the surface and inside the bead by monitoring the appearance of the ester carbonyl stretch at 1738 cm$^{-1}$. Plotting the intensity of this peak over time demonstrated that reactions in the interior and on the surface of a resin bead proceed at comparable rates.

The study of solid phase oxidation reactions made some interesting observations (Yan *et al.* 1996*b*). The IR spectrum of the Wang resin exhibited two prominent absorption bands in the hydroxyl stretching region: a sharp band at 3578 cm$^{-1}$ and a broad band at 3420 cm$^{-1}$. It was speculated that these could correspond with free hydroxyls and hydrogen bonded hydroxyl groups respectively. In the case of the hydrogen bonded groups it was not clear whether the hydrogen bonding was to other hydroxyls on the resin or to water molecules, although extensive heating and lyophilization did not reduce the intensity of the 3420 cm$^{-1}$ band. This study also demonstrated the power of IR microspectroscopy in optimizing conditions for solid phase synthesis. The rate of the TPAP catalysed oxidation of a benzylic alcohol could be determined by measuring the area under the hydroxyl and aldehyde carbonyl bands at various time points through the reaction. Comparing the oxidation with 0.2, 0.1, or 0.05 equivalents of TPAP indicated that the rate of reaction was reduced with the lesser amounts of catalyst.

An alternative approach for the quantitative analysis of solid phase chemical reactions depends on the observation of the IR carbon–deuterium stretching absorbance (Russell *et al.* 1996). The C–D absorption of deuterated compounds occurs between 2300–2200 cm$^{-1}$, a region generally free of other IR signals. To be able to quantitate the degree of chemical conversion in peptide synthesis the use of perdeuterated side-chain protecting groups in the

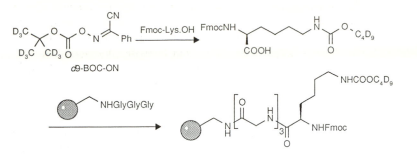

**Fig. 8.4** The preparation of a resin-bound peptide with perdeuterated Boc protection.

amino acids monomers was investigated. As the $d$9-Boc-protected amino acid coupled to the resin-bound peptide, the appearance of the C–D IR signal could be observed. In a similar fashion deprotection of the peptide chain N-terminus could be monitored by the disappearance of the C–D signal.

$d$9-BOC-ON was prepared and used to protect $\alpha$-Fmoc-lysine, and this *bis*-protected lysine was coupled to triglycine on aminomethyl cross-linked polystyrene resin (Fig. 8.4). The degree of chemical conversion was derived from the intensity of the C–D signal, negating any effect of the path length by making a comparison with the polystyrene backbone absorbance at $2600 \text{ cm}^{-1}$. Preparing samples containing $d$9-Boc groups in various positions on the peptides demonstrated that the different chemical environments had a negligible effect on the quantification of the deuterium, and that the technique was capable of 'counting' the number of d9-Boc–lysines in the resin-bound peptide.

One other infrared technique has been applied to the monitoring of solid phase reactions. Diffuse reflectance infrared Fourier transform spectroscopy (DRIFTS) is a valuable technique as it requires no sample preparation, has short analysis times, and lends itself to simple automation (Chan *et al.* 1997). As the resin beads do not need to be diluted with KBr, samples of a few milligrams only can be transferred directly to a metal cup at the focal point of the diffuse reflectance accessory, and spectra obtained in less than 30 seconds. Automation of the process permits up to 60 samples to be run successively. This method was used for monitoring the solid phase reduction of azido monosaccharides bound to TentaGel resin, observing the disappearance of the azide absorption at $2108 \text{ cm}^{-1}$.

## 3. Solid phase NMR

NMR spectroscopy is a familiar and accessible technique for organic chemists that provides a wealth of information that readily reveals chemical structures. The majority of NMR work is done on samples in solution, and thus the analysis of compounds whilst attached to resin beads presents new challenges for the technique. In particular the limited mobility of polymers themselves and the limited mobility of attached compounds leads to broad and poorly resolved signals. In addition the backbone may give rise to unwanted background signals that mask important detail in the compound spectra.

The NMR of resin-supported compounds however is not a new technique. As the solid phase synthesis of peptides was developed, a number of techniques had been developed to study synthetic progress. In particular, gel phase NMR spectra where the resin beads are swollen with solvent to maximize compound mobility had been used to examine both $^{19}$F and $^{13}$C nuclei. However, gel phase NMR is limited by the very low sensitivity of the small quantities of material on the bead. Large quantities, usually several hundred milligrams of resin were required, and acquisition times of hours were occasionally needed to generate spectra with an acceptable signal-to-noise ratio. More recently the advent of magic angle spinning (MAS) NMR has been successfully used for high-resolution $^{13}$C, proton, and correlation spectroscopy using much smaller quantities of resin and shorter acquisition times.

## 3.1 Carbon-13 NMR

Using broad band irradiation to remove coupling to protons, the $^{13}$C signals from the naturally abundant (1.1%) isomer of carbon can be observed as singlets in the gel phase NMR spectra. Polyacrylamide resin swollen with *d*6-DMSO was used for an early study of resins used in solid phase peptide synthesis (Epton *et al.* 1980). Irradiating 150 mg of resin with 30 000–150 000 pulses led to signals being observed for nearly all of the groups attached to the polymer backbone. These could be readily assigned by reference to the NMR spectra of the polymer precursor monomers. Attaching Boc–leucine to the resin beads and removal of the Boc protecting group using boron trifluoride–etherate could be followed by NMR. It was observed that after deprotection, a $^{13}$C signal corresponding to the Boc methyl groups still remained, indicating that around 11% of the protecting groups had not been removed. This was an important result. For the first time, $^{13}$C NMR spectroscopy had been used to show that a reaction had not reached completion, and that further acid treatment was required.

Solid phase peptide synthesis on a range of cross-linked polystyrene resin beads has also been monitored by $^{13}$C gel phase NMR (Giralt *et al.* 1984). The resin swollen with an apolar solvent such as deuterochloroform or dioxane generated spectra of the polymer backbone. Because of the potential for complex backbone relative stereochemistry (heterotacticity), most of the carbon signals were observed to have a fine multiplicity of structure. However, the *p*-carbon of the phenyl ring was least affected by the backbone configuration and appeared as a singlet. It was thus assumed that the signals derived from attached molecules would also not be affected by the polymer structure avoiding complications of fine structure.

Merrifield, Wang, and benzhydrylamine resins were characterized by gel phase NMR spectra. As expected the chloromethyl signal of Merrifield resin was a narrow singlet observed at 46.3 p.p.m., a result consistent with distance from the backbone leading to significant molecular mobility. A weak signal at 65.3 p.p.m. was observed, and assumed to represent contaminating hydroxymethyl resin within the chloromethyl resin. Wang resin produced narrow signals for the aromatic carbons indicating that pendant aromatic signals are easier to observe than substituted phenyl rings on the polymer backbone. Synthesizing peptides on polystyrene resin was readily monitored by $^{13}$C NMR. A study of the relaxation times ($T_1$) of the carbons in resin-bound

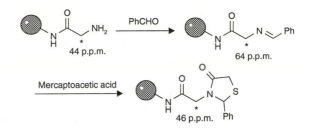

**Fig. 8.5**   The synthesis of a thiazolidinone on TentaGel resin was followed by gel phase [13]C NMR. The [13]C-enriched atoms are marked (*) with the observed chemical shifts.

peptides indicates the relative mobility in the molecules. Indeed, measuring the $T_1$ values of the $\alpha$-carbon along the peptide chain suggested that there was no significant mobility gradient.

The major problem with [13]C NMR is the low natural occurrence of this carbon isotope. Several thousand transients need to be acquired on tens of milligrams of resin to give spectra with reasonable signal-to-noise ratios. Fortunately, the speed of spectra determination can be accelerated by the use of [13]C-enriched substrates on the solid phase (Look *et al.* 1994) making this a practical technique for reaction monitoring. $\alpha$-[13]C-enriched glycine was attached to TentaGel resin, and the free amine used in a thiazolidinone synthesis. By employing the enriched starting material, only this signal (plus a signal due to the polyethylene glycol resin) was observed, and chemical conversion of the glycine was reflected in distinctive chemical shift changes. Using enriched materials allowed much briefer NMR experiments; spectra were recorded with only 64 transients, reducing the turnaround time to between 15 and 30 minutes.

Converting TentaGel-bound glycine to an imine was mirrored by a change in the glycine $\alpha$-carbon chemical shift from 44 p.p.m. to 64 p.p.m. (Fig. 8.5). Condensation of the resin-bound imine with mercaptoacetic acid gave the thiazolidinone. The [13]C signal moved from 64 p.p.m. to 46 p.p.m. during this synthetic sequence. Any incomplete reaction would be evident from a residual signal at the original chemical shift. In all cases the observed chemical shifts were consistent with the shifts of the corresponding solution analogues.

In a similar manner, the use of [13]C-enriched benzaldehyde, allowed the progress of the synthesis of pyridines and pyrimidines to be followed using gel phase NMR (Gordeev *et al.* 1996).

[13]C gel phase NMR spectroscopy has been used to monitor the progress of solid supported reactions in 'real-time' (Barn *et al.* 1996). The alkylation of amines with [13]C-enriched bromoacetic acid attached to both Wang and TentaGel resins was carried out in deuterated DMF in an NMR tube. The labelled methylene group of the starting material at 27.4 p.p.m. could be seen after just ten minutes acquisition time, and following the addition of piperidine a new signal was rapidly observed at 60.3 p.p.m. The reaction was shown to proceed to completion in just 30 minutes.

Overall, gel phase [13]C NMR gives relatively narrow signals and variations in mobility successfully separate product signals from the polymer resonances. As described above, the technique has been used successfully to monitor the progression of solid phase reactions, but the poor sensitivity restricts the speed with which spectra can be generated. Although $^{13}$C-enriched material has been used to accelerate acquisition, the advent of MAS NMR has allowed more rapid spectroscopy with smaller resin samples (see Section 3.4).

## 3.2 Fluorine-19 NMR

Following the success of $^{13}$C NMR spectra of solvent swollen cross-linked polystyrene, the spectra of $^{19}$F labelled compounds attached to solid phase have been investigated. As a naturally occurring nuclei, $^{19}$F offers obvious advantages over $^{13}$C, avoiding the need for enriching compounds or lengthy acquisition times to overcome low natural $^{13}$C abundance. However, fluorine is an uncommon element in organic synthesis, and thus to use $^{19}$F NMR to monitor solid phase synthesis requires the use of specific fluorinated intermediates or reactions.

Peptide synthesis has been carried out on Merrifield resin using fluorine-containing protecting groups (either fluorobenzylcarbonyl or 2-(4-fluorophenyl)-2-propyloxycarbonyl) (Manatt *et al.* 1980). The amino acid derivatives exhibited proton-decoupled singlet $^{19}$F NMR signals characteristic of aryl fluorides and the chemical shifts could be used in a diagnostic fashion for a range of amino acid–resin systems. More recently, the progress of a chemical reaction has been monitored by the observation of the fluorine NMR signal (Shapiro *et al.* 1996). The aromatic nucleophilic substitution reaction involving the displacement of fluorine by an amine nucleophile was observed on Rink resin by gel phase $^{19}$F NMR (Fig. 8.6). Plotting the reaction by acquiring spectra at various time points demonstrated the reaction was complete within two hours. $^{19}$F MAS NMR taken at several time points gave essentially the same message although these spectra took four minutes to acquire, compared with the 1–1.5 hours required for each gel phase spectrum.

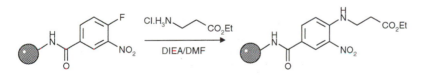

**Fig. 8.6** The nucleophilic displacement of an aromatic fluoride was monitored by $^{19}$F NMR.

Reactions on TentaGel resin beads have also been monitored using gel phase $^{19}$F NMR (Svensson *et al.* 1996). The $^{19}$F chemical shift of *o*-, *m*-, and *p*-fluorobenzoic acid attached to TentaGel amine resin are very similar to the shifts observed from the corresponding 2-methoxyethylbenzamides in solution, indicating the similarity of environment between solution and TentaGel-bound compounds (Fig. 8.7). The gel phase spectra were obtained in less than ten minutes using 50–200 mg of resin, and the linewidths observed approached those of solution spectra. This approach was used to monitor a synthetic sequence involving an aryl fluoride, and TentaGel permitted the generation of high quality spectra without the need to resort to MAS NMR.

## 3.3 Phosphorus-31 NMR

The natural abundance of the $^{31}$P nuclei makes it another obvious choice for the NMR monitoring of solid phase reactions. Gel phase $^{31}$P NMR has been

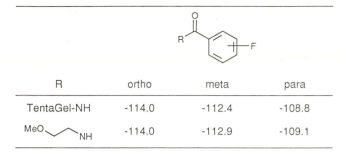

| R | ortho | meta | para |
|---|---|---|---|
| TentaGel-NH | -114.0 | -112.4 | -108.8 |
| MeO-NH | -114.0 | -112.9 | -109.1 |

**Fig. 8.7**  The $^{19}$F NMR chemical shifts of fluorobenzamides attached to TentaGel resin and in solution.

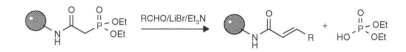

**Fig. 8.8**  The Horner–Wadsworth–Emmons reaction monitored by $^{31}$P gel phase NMR.

used (Johnson and Zhang 1995) to study the Horner–Wadsworth–Emmons synthesis of olefins from resin-bound diethylphosphonoacetamide and aldehydes (Fig. 8.8). The NMR of the starting material showed the phosphorus as a narrow multiplet at $\delta$ 22, but this disappears during the reaction to be replaced by the signal of diethyl phosphate near $\delta$ 0.

In this example the gel phase NMR was facilitated by the use of polyethylene glycol grafted resin—the flexibility of the chains providing a mobile environment for the molecules being examined.

## 3.4  Magic angle spinning NMR

Proton NMR has the capacity to convey a wealth of structural information from solution samples, and is thus a linchpin of synthetic organic chemistry. However, when applied to the analysis of samples attached to resin beads, the spectra are much less informative. Although proton NMR offers sufficient sensitivity to be used for the analysis of solid phase reactions, this approach is generally compromised by broad and featureless spectra resulting from either restricted molecular motion or magnetic inhomogeneities surrounding the resin bead.

MAS NMR is a technique that is widely used for solid state samples to remove line broadening. Although it has generally been applied to solids in the past, MAS NMR has also found utility in the analysis of very small liquid samples. Consequently, it has been found that high-resolution proton NMR for solid phase synthesis could be achieved by the use of MAS and a nanoprobe specifically designed for very small (< 40 $\mu$L) liquid samples (Fitch *et al.* 1994). This probe has a highly effective detection system as 100% of the sample is placed within the receiver coil. For example, a solid phase synthesis product (**8.1**) attached to TentaGel resin was found to give narrow line spectra when analysed by this method (Fig. 8.9). Approximately 10 mg of resin, containing 1.5 mg of product, was suspended in 30 $\mu$L of *d*6-DMSO, and only 16 transients were necessary to give a spectrum that had a threefold improvement in signal-to-noise ratio compared with a conventional proton spectrum obtained on a 100 mg resin sample.

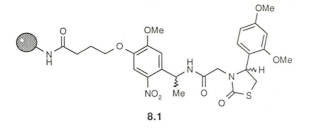

**8.1**

The nanoprobe has also been employed to obtain an NMR spectrum from a single bead (Sarkar *et al.* 1996). To avoid the signal from the product being complicated by signals from the Wang resin polymer backbone, the compound was $^{13}$C labelled and an isotope-filtered NMR process was used to observe only the protons attached to the $^{13}$C atoms. Although a successful experiment highlighting the sensitivity of the nanoprobe, the use of an isotopically labelled substrate limits this approach for general solid phase combinatorial chemistry.

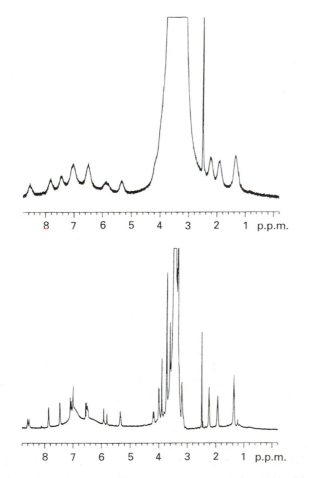

**Fig. 8.9** (*Top*) A 16-scan 500 MHz NMR spectrum of **8.1**. (*Bottom*) A 16-scan 500 MHz NMR spectrum of **8.1** obtained by spinning at 2000 Hz around the magic angle. Reprinted with permission from *J. Org. Chem.* (1994), **59**, 7955-6. Copyright 1994 American Chemical Society.

Judicious choice of resin and solvent has a highly beneficial effect on the quality of MAS proton spectra (Keifer 1996). In this study, a range of resins bearing the same *t*-butyl aspartic acid were examined. Sensitivity is not a limiting issue for the nanoprobe, as it can generate spectra even from a single bead, but the quality of spectra is dictated solely by linewidth. TentaGel, by virtue of the long linking polyethylene glycol chains, produced narrow linewidths in the greatest range of solvents. Wang and Rink resins with their shorter linkers and thus more restricted mobility occasionally produced acceptable spectra, but polystyrene resins that hold the compound very close to the backbone such as Merrifield never generated high quality spectra. Solvents also affected the quality of spectral data. *d*7-DMF, *d*2-dichloro-methane, and *d*6-DMSO typically produced the best spectra. Acetone, methanol, and benzene were sometimes useful, but $D_2O$ was never suitable.

A similar result was obtained from another study of four different solid supports (Wehler and Westman 1996). TentaGel possessed the best NMR properties, but other resins including Merrifield polystyrene were found suitable for NMR analysis.

In addition to the desired signals from the bound compound, the beads also generate a range of signals originating from the resin backbone, and although these can be partly removed by pre-saturation, the residual signals can mask or complicate spectral assignments. Recent developments in pulse sequences can almost totally remove the polymer resonances (Shapiro *et al.* 1997) but also results in the loss of coupling information. The use of J-resolved 2D NMR, and projecting the non-tilted spectrum onto the chemical shift dimension gives enhanced resolution and removal of the complicating polymer signals.

MAS NMR also permits the generation of good quality 2D NMR spectra allowing the total structural determination of a compound whilst attached to a solid support. Total correlation spectroscopy (TOCSY) NMR spectrum of Fmoc–Lys–Boc attached to Wang resin (Anderson *et al.* 1995), and COSY and TOCSY spectra of a lithium aluminium hydride reduction product on Merrifield resin (Garigipati *et al.* 1996) have been obtained.

All of the above examples refer to the use of MAS NMR for the analysis of compounds attached to resin beads. However, many other solid supports have been used for library synthesis. Of these, the multipin approach is a popular way to prepare libraries of single compounds. It has now been shown that it is possible to analyse the chemical structures of compounds attached to multipin crowns (Chin *et al.* 1997). The entire crown was placed in *d*7-DMF, and proton NMR spectra obtained using a spin lock time in the pulse sequence that largely removed the broader signals due to the plastic support. It was possible to observe the disappearance of an aldehydic signal at $\delta = 10.17$ p.p.m., and the appearance of olefinic protons through a Wittig reaction (Fig. 8.10). A COSY experiment was also performed to demonstrate the *trans* stereo-chemistry of the olefin bond. One advantage of this method was that the crown could be returned to the reaction vessel after NMR analysis should the reaction be incomplete.

Related to synthesis on multipin crowns is the development of single macro beads for solid phase chemistry. These beads, which are 400–750 $\mu$m diameter polyethylene glycol-polystyrene copolymers and can hold 10–65 nmol product per bead, can also be analysed singly by MAS NMR (Pursch *et al.* 1996). A single bead, swollen in a solvent such as *d*6-DMSO or $CDCl_3$, was placed in an

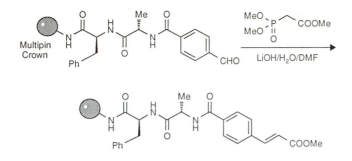

**Fig. 8.10** The Wittig reaction of an aldehyde attached to a Multipin crown and analysed by proton MAS NMR.

NMR rotor, and the proton NMR readily obtained, thus allowing the monitoring of a synthetic sequence to give hydantoin products.

The demand for solid phase chemistry to meet the appetites of combinatorial chemists has forced the pace in the arena of solid phase NMR techniques. The use of MAS NMR spectroscopy and pulse sequences that remove the worst of the complicating polymer resonances now permit spectra of solution quality to be generated. Whether these will be used routinely in the analysis of combinatorial library compounds will become evident over the next few years.

## 4. Mass spectrometry

Unlike IR and NMR, MS is a destructive technique that necessarily requires cleavage of the compound being examined from the solid phase. However, it is an exquisitely sensitive technique, able to produce data from picomole to femtamole quantities of compound. Three MS techniques have been employed so far in the analysis of combinatorial library samples. These are matrix-assisted laser desorption/ionization time-of-flight MS (MALDI–TOF MS), imaging time-of-flight secondary ion MS (TOF–SIMS) and electrospray ionization MS (ESI-MS). The use of MS in combinatorial chemistry falls into two main categories:

- the monitoring of solid phase products and reactions
- the identification of active compounds from libraries

### 4.1 The analysis of solid phase chemistry

MALDI–TOF MS is a new MS technique that uses laser energy to desorp and ionize involatile molecules, such as peptides, intact from a matrix. High sensitivity allows the detection of femtamole compound quantities—an attribute that is essential if examining single beads containing low quantities of material. The technique was used to examine an oligomeric N-substituted glycine ('peptoid') synthesized on Rink resin (Zambias *et al.* 1994). A single resin bead containing the compound was placed in a stream of TFA vapour for five minutes. In this way the product was cleaved from the support but could not diffuse away from the bead. The bead was attached to the MS sample target and a solution of the dihydroxybenzoic acid matrix (used to assist the desorption of large and involatile molecules) was allowed to crystallize around the bead. Irradiation with a 337 nm laser beam generated ions of the parent peptoid in the MS confirming the molecular weight of the parent compound. Some fragmentation of the peptoid was observed giving additional structural information.

Peptides on single or groups of beads have been examined using the same technique (Egner *et al.* 1995*a*). A number of peptides were synthesized and analysed using this technique and in each case the protonated or sodiated molecular ions were clearly observed. The observation of unexpected ions permitted the resolution of synthetic problems that would otherwise have been undetected. The same group have examined compounds attached to other resin types (Egner *et al.* 1995*b*). The preparation of resin samples for MS involved cleaving Wang and HMP linkers with TFA to give carboxylic acids, or 2-chlorotrityl resin cleaved with TFA to give amines. Base-labile linkers were cleaved with ammonia vapour and irradiation cleaved light-sensitive linkers. Consistent with the habit of mass spectroscopists to assign acronyms to new techniques, this approach has been named SPIMS, solid phase *in situ* mass spectrometry, reflecting the fact that the beads are not removed prior to analysis.

TOF–SIMS offers mass accuracy of the order of ±0.01 atomic mass units, and thus can be used to distinguish between products of the same nominal molecular mass. In particular this approach allows the ready identification of active peptides in a large combinatorial library by the examination of a single bead bearing the active compound (Brummel *et al.* 1994). In the cases where isomeric sequences cannot be distinguished by accurate mass alone, the secondary ions produced by fragmentation can unequivocally determine the peptide sequence. For example, the tripeptide Val–Tyr–Val was cleaved from the bead before being subjected to TOF–SIMS, and in addition to ions for the tripeptide, signals corresponding to Val–COOH and Tyr–COOH and the dipeptides were observed.

Capillary electrophoresis is a solution-based technique that allows the rapid and efficient separation of minute quantities of material. Separations are based on the relative electrophoretic mobilities of charged species in an electric field within a small diameter column of fused silica. Coupled with mass spectrometry to characterize the separated components, it offers a powerful method for combinatorial mixture analysis.

CE in combination with electrospray ionization mass spectrometry (CE–ESI MS) has been used to determine the purity of a library mixture (see Chapter 4, Section 4) of 171 disubstituted xanthene derivatives (Dunayevskiy *et al.* 1996). Even though 124 of the compounds were isobaric (i.e. have the same molecular weight) with at least one other compound within the mixture, CE allowed separation of these isomers and overall 160 of the compounds were shown to be present.

## 4.2   The identification of active library components

In addition to its use in checking the integrity of combinatorial library mixtures, CE has been shown to be useful in the measurement of binding constants. Bringing these various factors together with MS allows the creation of a method for the identification of compounds from within a combinatorial mixture that preferentially binds to a target protein. This approach, affinity capillary electrophoresis mass spectrometry (ACE MS) has been used to find peptides that bind to vancomycin, a naturally occurring antibiotic that works as a soluble peptide receptor (Chu *et al.* 1995).

A peptide library of 100 components of the structure, Fmoc–DDXX (where X was a mixture of ten possible amino acids) was injected as the sample into the CE column. Vancomycin was present within the buffered solution, and peptides that bound to the compound underwent multiple association–dissociation events. Through this process of complexation to the receptor, the mobility of the ligand peptides shifts towards that of the complex. These ligands were retained relative to the other peptides in the mixture and were eluted and detected later. In this manner, three tight vancomycin-binding peptides were detected and ESI MS identified these as Fmoc–DDFA, Fmoc–DDYA, and Fmoc–DDHA. A subsequent publication from the same group has extended ACE MS to mixtures of up to 1000 peptides (Chu *et al.* 1996) with a greater range of amino acids in the component peptides.

The binding of peptide library components to enzymes can also be examined by ESI-MS without the use of a chromatographic separation. Two libraries of dipeptide derivatives (289 and 256 components) of *p*-carboxy-benzene sulfonamide (**8.2**) were screened for inhibitory activity against carbonic anhydrase II (Gao *et al.* 1996). This study was carried out by generating ions of the non-covalently bound complex in the gas phase and using the technique of Fourier transform ion cyclotron resonance (FTICR) to identify the tight binding inhibitors after dissociation from the enzyme. The ligand with the dipeptide sequence X–X = L-Leu-L-Leu with a binding constant of 14 nM was discovered.

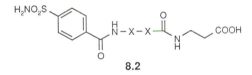

**8.2**

The above examples demonstrate that MS is a highly varied set of techniques that have a huge potential in the analysis of combinatorial libraries. In addition to the straightforward conformation that a desired compound is actually present on a single resin bead from a solid phase synthesis, MS techniques in combination with CE will resolve library mixtures of high complexity. The latest and perhaps the most exciting application of MS is the development of techniques that allow screening of the library mixture and identification of potent ligands within the MS system itself. This rapidly growing field will increasingly impact on combinatorial library design, analysis and assessment in the future.

# References

Anderson, R.C., Stokes, J.P., and Shapiro, M.J. (1995). Structural determination in combinatorial chemistry: utilization of magic angle spinning HMQC and TOCSY NMR spectra in the structural determination of Wang-bound lysine. *Tetrahedron Lett.*, **36**, 5311–14.

Barn, D.R., Morphy, J.R., and Rees, D.C. (1996). Synthesis of an array of amides by aluminium chloride assisted cleavage of resin-bound esters. *Tetrahedron Lett.*, **37**, 3212–16.

Brummel, C.L., Lee, I.N.W., Zhou, Y., Benkovic, S.J., and Winograd, N. (1994). A mass spectrometric solution to the address problem of combinatorial libraries. *Science*, **264**, 399–402.

Chan, T.Y., Chen, R., Sofia, M.J., Smith, B.C., and Glennon, D. (1997). High throughput on-bead monitoring of solid phase reactions by diffusion reflectance infrared Fourier transform spectroscopy (DRIFTS). *Tetrahedron Lett.*, **38**, 2821–4.

Chin, J., Fell, B., Shapiro, M.J., Tomesch, J., Wareing, J.R., and Bray, A.M. (1997). Magic angle spinning NMR for reaction monitoring and structure determination of molecules attached to Multipin crowns. *J. Org. Chem.*, **62**, 538–9.

Chu, S.S. and Reich, S.H. (1995). NPIT: A new reagent for quantitatively monitoring reactions of amines in combinatorial synthesis. *Bioorg. Med. Chem. Lett.*, **5**, 1053–58.

Chu, Y-H., Kirby, D.P., and Karger, B.L. (1995). Free solution identification of candidate peptides from combinatorial libraries by affinity capillary electrophoresis/mass spectrometry. *J. Am. Chem. Soc.*, **117**, 5419–20.

Chu, Y-H., Dunayevskiy, Y.M., Kirby, D.P., Vouros, P., and Karger, B.L. (1996). Affinity-capillary electrophoresis-mass spectrometry for screening combinatorial libraries. *J. Am. Chem. Soc.*, **118**, 7827–35.

Dunayevskiy, Y.M., Vouros, P., Wintner, E.A., Shipps, G.W., Carell, T., and Rebek, J. (1996). Application of capillary electrophoresis–electrospray ionization mass spectrometry in the determination of molecular diversity. *Proc. Natl. Acad. Sci. USA*, **93**, 6152–7.

Egner, B.J., Langley, G.J., and Bradley, M. (1995*a*). Solid phase chemistry: direct monitoring by matrix-assisted laser desorption/ionization time of flight mass spectrometry. A tool for combinatorial chemistry. *J. Org. Chem.*, **60**, 2652–3.

Egner, B.J., Cardno, M., and Bradley, M. (1995*b*). Linkers for combinatorial chemistry and reaction analysis using solid-phase *in situ* mass spectrometry. *J. Chem. Soc. Chem Commun.*

Epton, R., Goddard, P., and Ivin, K.J. (1980). Gel phase 13C n.m.r. spectroscopy as an analytical method in solid (gel) phase peptide synthesis. *Polymer Commun.*, **21**, 1367–71.

Fitch, W.L., Detre, G., Holmes, C.P., Shoolery, J.N., and Keifer, P.A. (1994). High-resolution 1H NMR in solid-phase organic synthesis. *J. Org. Chem.*, **59**, 7955–6.

Frechet, J.M. and Schuerch, C. (1971). Solid-phase synthesis of oligosaccharides. I. Preparation of the solid support. Poly[*p*-(1-propen-3-ol-1-yl)styrene]. *J. Am. Chem. Soc.*, **93**, 492–6.

Gao, J., Cheng, X., Chen, R., Sigal, G.B., Bruce, J.E., Schwartz, B.L., *et al.* (1996). Screening derivatized peptide libraries for tight binding inhibitors to carbonic anhydrase II by electrospray ionization–mass spectrometry. *J. Med. Chem.*, **39**, 1949–55.

Garigipati, R.S., Adams, B., Adams, J.L., and Sarkar, S.K. (1996). Use of spin echo magic angle spinning $^1$H NMR in reaction monitoring in combinatorial organic synthesis. *J. Org. Chem.*, **61**, 2911–14.

Giralt, E., Rizo, J., and Peroso, E. (1984). Application of gel-phase $^{13}$C-NMR to monitor solid phase peptide synthesis. *Tetrahedron*, **40**, 4141–52.

Gisin, B.F. (1972). The monitoring of reactions in solid-phase peptide synthesis with picric acid. *Anal. Chim. Acta*, **58**, 248–9.

Gordeev, M.F., Patel, D.V., Wu, J., and Gordon, E.M. (1996). Approaches to combinatorial synthesis of heterocycles: solid phase synthesis of pyridines and pyrido[2,3-d]pyrimidines. *Tetrahedron Lett.*, **37**, 4643–6.

Hancock, W.S. and Battersby, J.E. (1976). A new micro-test for the detection of incomplete coupling reactions in solid-phase peptide synthesis using 2,4,6-trinitrobenzenesulphonic acid. *Anal. Biochem.*, **71**, 260–4.

Johnson, C.R. and Zhang, B. (1995). Solid phase synthesis of alkenes using the Horner–Wadsworth–Emmons reaction and monitoring by gel phase $^{31}$P NMR. *Tetrahedron Lett.*, **36**, 9253–6.

Keifer, P.A. (1996). Influence of resin structure, tether length, and solvent upon the high-resolution 1H NMR spectra of solid-phase-synthesis resins. *J. Org. Chem.*, **61**, 1558–9.

Look, G.C., Holmes, C.P., Chinn, J.P., and Gallop, M.A. (1994). Methods for combinatorial organic synthesis: the use of fast 13C NMR analysis for gel phase reaction monitoring. *J. Org. Chem.*, **59**, 7588–90.

Manatt, S.L., Amsden, C.F., Bettison, C.A., Frazer, W.T., Gudman, J.T., Lenk, B.E., *et al.* (1980). A fluorine-19 NMR approach for studying Merrifield solid-phase peptide synthesis. *Tetrahedron Lett.*, **21**, 1397–400.

Pursch, M., Schlotterbeck, G., Tseng, L-H., Albert, K., and Rapp, W. (1996). Monitoring the reaction progress in combinatorial chemistry: $^1$H MAS NMR investigations on single macro beads in the suspended state. *Angew. Chem. Int. Ed. Engl.*, **35**, 2867–9.

Reddy, M.P. and Voelkner, P.J. (1988). Novel method for monitoring the coupling efficiency in solid phase peptide synthesis. *Int. J. Pept. Protein Res.*, **31**, 345–8.

Russell, K., Cole, D.C., McLaren, F.M., and Pivonka, D.E. (1996). Analytical techniques for combinatorial chemistry: quantitative infrared spectroscopic measurements of deuterium-labeled protecting groups. *J. Am. Chem. Soc.*, **118**, 7941–5.

Sarin, V.K., Kent, S.B.H., Tam, J.P., and Merrifield, R.B. (1981). Quantitative monitoring of solid-phase peptide synthesis by the ninhydrin reaction. *Anal. Biochem.*, **117**, 147–57.

for drug discovery are the pharmaceutical companies or associated biotechnology companies.

One of the earliest uses of combinatorial synthesis for drug discovery was the synthesis of a peptide library described in a paper from the Lilly Research Laboratories (Owens *et al.* 1991). The process of mix and split synthesis followed by iterative deconvolution was used for the discovery of a potent pentapeptide inhibitor of the enzyme, HIV protease. Initially, a library of over 240 000 acylated tetrapeptides were synthesized in 22 mixtures of the form; Ac–$D_4$–$X_3$–$Z_2$–$X_1$–$NH_2$, where $X_1$ and $X_3$ are mixtures of 22 diverse amino acids (including both L- and D- stereochemistries), $Z_2$ is the same mixtures with the addition of statine, and $D_4$ is a defined amino acid residue. Iterative deconvolution of the most active mixtures at each round revealed the potent sequence: Ac–Phe–Ile–Sta–D-Leu–$NH_2$ ($IC_{50} = 1.4$ $\mu$M). The presence of statine in the final compounds was expected as this amino acid is key pharmacophore for the inhibition of aspartyl proteases, the class of enzyme of which HIV protease is a member. This tetrapeptide served as a starting point for further investigation, leading to the discovery of a more active pentapeptide, Ac–Trp–Val–Sta–D-Leu–$X_1$–$NH_2$ ($X_1$ not defined in the paper), with an $IC_{50}$ value of 50 nM.

The mix and split approach has also been used by a group at Pfizer for the synthesis of a 31 000 component library of tripeptides prepared to identify potent endothelin antagonists and derive limited SAR around a previously reported Fujisawa endothelin inhibitor, FR-139317 (**9.1**) (Terrett *et al.* 1995). A pool of 32 diverse Fmoc-protected amino acids, that included only seven naturally occurring $\alpha$-amino acids, was used to assemble the library in 32 mixtures of 992 triamides using a standard solid phase coupling methodology (TBTU–HOBT–DIPEA). Screening and iterative deconvolution identified FR-139 317 as the most active library component and highlighted several potent analogues including **9.2**.

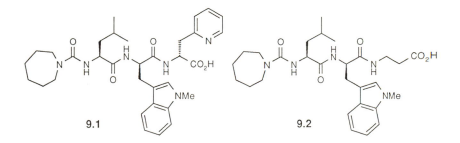

9.1                                        9.2

Proteases responsible for the hydrolysis of specific peptides or proteins will have a preferred substrate specificity. A knowledge of this substrate can be useful in understanding the enzyme mechanism and furthermore in the design of potent and selective inhibitors. The synthesis and screening of combinatorial peptide libraries is an effective way to identify this specificity. For example, positional scanning peptide libraries (see Chapter 2, Section 6) have been used to define the preferred substrate specificity of interleukin-1$\beta$ converting enzyme (ICE) (Rano *et al.* 1997). A library of peptides of the form Ac–X–X–X–Asp–aminomethylcoumarin were prepared as potential substrates of the enzyme. This structure was chosen as there was known to be a requirement for aspartic acid on the N-terminal side of the peptide immediately adjacent to the bond hydrolysed. There was also a requirement for substrates of at least four residues to be recognized by the enzyme, and the aminomethylcoumarin was included to provide a fluorescent leaving group that could be determined in the assay. Preparing the library in three sublibraries of 20 mixtures of 400 compounds allowed the investigation of the preference for each residue in the sequence. Following the positional scanning experiment, the optimal peptide substrate was determined to be WEHD, a tetrapeptide sequence recognized some 50-fold more efficiently than the natural substrate sequence, YVAD. This result indicates that the enzyme may possibly be responsible for the hydrolysis of additional endogenous substrates.

One of the most exciting aspects of this work was the manner in which knowledge of the substrate sequence immediately led to the design of a potent ICE inhibitor. Replacing the scissile bond of the substrate with an aldehyde, a functionality that binds well to the nucleophilic cysteine thiol in the active site of ICE, gave Ac–WEHD–CHO. This inhibitor had extremely high affinity for the enzyme, with a $K_i$ value of 56 pM.

Progressing beyond the synthesis of peptide libraries, many drug discovery programmes have benefited from the use of combinatorial chemistry to prepare libraries of non-peptides. This book has already described the innovative synthesis of *N*-substituted glycines ('peptoids') investigated by Chiron (see Chapter 6, section 5), and the synthesis of ACE inhibitor proline derivatives by Affymax (see Chapter 5, Section 3.4 and Chapter 7, Section 3.4). In each of these cases very potent receptor ligands or enzyme inhibitors were discovered. A few other examples of successful combinatorial chemistry applications to the process of drug discovery deserve special mention.

In recent years there have been a large number of groups seeking to discover inhibitors of HIV protease, a key enzyme involved in the processing of the HIV-1 viral proteins. It has been discovered that compounds possessing a $C_2$ symmetrical diamino diol or pseudosymmetrical diamino alcohol core constitute a particularly potent series of inhibitors. This information was used by Abbott in the design and synthesis of a library of compounds destined for an HIV protease screen (Wang *et al.* 1995). In fact, by linking either the alcohol or diol to an MBHA-derivatized solid phase, the two amines could be simultaneously functionalized to build symmetrical molecules, a process which differs from the usual linear solid phase route to compounds such as peptides. Using the Fmoc-protection strategy, each of the two core molecules was derivatized with 150 different acylating and sulfonylating groups to give a total library of 300 individual compounds (Fig. 9.1). Screening against HIV protease revealed a number of active inhibitors; some with $IC_{50}$ values below 100 nM.

The impact of combinatorial chemistry on drug discovery is now becoming apparent through reports of clinical candidates discovered using these high speed techniques. Chapter 1 highlighted the distinction between the use of library synthesis for lead discovery and lead optimization. In the former case, the techniques of combinatorial chemistry are used to generate large numbers of diverse compounds in the hope of discovering previously unknown and unprecedented leads. With that lead in hand, smaller and less diverse arrays of single compounds can be assembled to optimize the potency or selectivity of the lead structure. Of these two applications, the latter approach is more likely

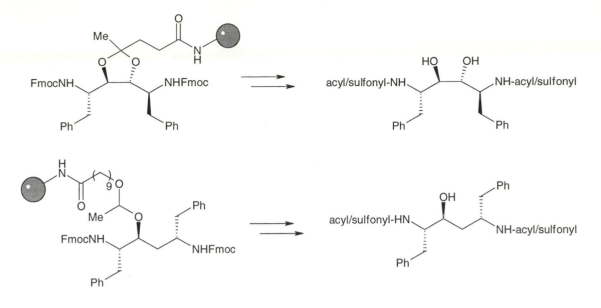

**Fig. 9.1**   The linking of diamino diol and diamino alcohol cores to solid phase and the generation of analogues for screening against HIV protease.

to give success, as the compounds are designed around a validated biologically active lead compound, rather than searching for a compound with affinity for a previously unexplored class of receptor or enzyme. The successes now being reported tend to be those compounds that build on a well known structural series. The American pharmaceutical company Lilly have a compound currently in clinical development for the therapy of migraine (Kaldor 1997). The remarkable speed of development saw the discovery programme advance from identification of the receptor subtype to early clinical trials in man in under two years. The compound, LY334370 (**9.3**) is an antagonist of 5HT1F, a newly identified subclass of a well-known family of serotonin receptors. The synthetic programme synthesized around 500 analogues of serotonin, and LY334370 emerged as the best of these compounds—a clear example of accelerating the process of lead optimization through rapid parallel synthesis.

Merck have reported a similar drug discovery success story through the combinatorial synthesis of analogues. In this case the target was a compound that would activate the release of growth hormone, and the analogues synthesized were based on a proprietary compound (Patchett *et al.* 1995). The design breakthrough in this work was provided by the compound **9.4**, a derivative of tryptophan that readily lent itself to the rapid synthesis of analogues. The preparation of numerous analogues led to the identification of L-163191 (MK-0677, **9.5**), a compound with a 40-fold improvement in activity over **9.4** and currently in late stage clinical development.

These are two examples only of a growing family of successful drug discoveries that owe their existence to the use of some form of library chemistry. Although in both cases it seems that combinatorial chemistry was used for lead optimization, it is only a matter of time before drug molecules

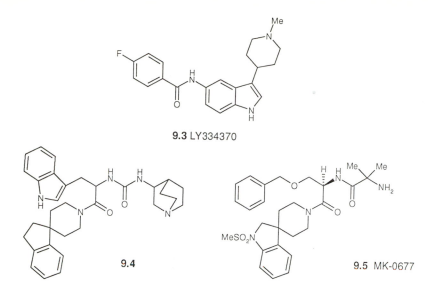

**9.3** LY334370

**9.4**

**9.5** MK-0677

are described that were identified through programmes that started with random screening of a large combinatorial library. We can confidently expect that combinatorial chemistry will facilitate the discovery of totally new molecules and pharmacological activities, at a rate previously unknown. Novel classes of structure that may be as significant as the dihydropyridines, β-lactams, or benzodiazepines could be discovered and take medicinal chemistry in new directions, producing highly effective and safe drug molecules for the next century.

## 2.   Virtual libraries and combinatorial chemistry

One concept that is widely used by combinatorial chemists is that of the virtual library (VL). As the name implies, this is a set of compounds that in reality does not exist, but could be generated if desired using known chemistry and available monomeric building blocks. The VL concept is important because it is possible to use an electronic database to generate and store the VL, and use computational searching tools to select which compounds should be synthesized. The selection tools are many and varied, but could include calculating physicochemical properties of the molecules such as lipophilicity, molecular weight, or dipole moment. Alternatively, a diversity algorithm could be applied to select a set of compounds as different from each other as possible. If an active compound had been discovered, similarity measures might be applied to find other compounds from the VL with possibly similar or better biological activity.

The use of virtual libraries in the discovery and optimization of a cathepsin D inhibitor has been described (Kick *et al.*, 1997). Cathepsin D is an aspartyl protease that has been implicated in tumour metastasis in breast cancer, melanoma metastasis, and Alzheimer's disease. An X-ray crystallographic structure of the enzyme with the natural inhibitor pepstatin complexed to the active site was the starting point for the design of new compounds that might be potent inhibitors. Aspartyl proteases are known to be inhibited by

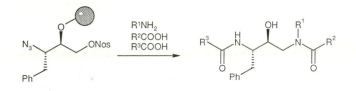

**Fig. 9.2** The synthesis of a library of cathepsin D inhibitors from a resin-linked (hydroxyethyl)amine template and three positions of diversity.

substituted (hydroxyethyl)-amines, as these templates mimic the enzyme hydrolysis transition state. This template was thus used as the focus for library design, introducing structural variation in three positions (Fig. 9.2).

The VL was generated from lists of 700 amines and around 1900 acylating and sulfonylating agents. A virtual combinatorial library of all these building blocks would contain over one billion possible compounds. To reduce this number to a manageable size, a computational method entitled CombiBuild was used to generate three sets of ten monomers chosen on the basis of fit to the X-ray structure following conformational analysis. In a highly pragmatic fashion, the monomer price was also used as a criterion for selection! Those monomers were then used to prepare a 'directed' combinatorial library of 1000 individual analogues on polystyrene bead solid support. For comparative purposes a 'diverse' set of 1000 compounds were also selected without using structure-based design methods. Both sets were tested for inhibitory activity against cathepsin D leading to seven compounds from the 'directed' library with an $IC_{50}$ below 100 nM compared with just one compound of similar activity from the 'diverse' library.

Following the discovery of active lead compounds, structural optimization led to the design of a second generation library of 39 compounds, picking monomers related to those that earlier had yielded the active enzyme inhibitors monomers. This process gave rise to compounds with a five- to sixfold improvement in potency, including the inhibitor **9.6** ($IC_{50} = 14$ nM).

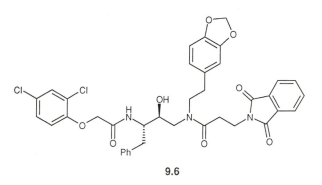

**9.6**

Overall, this study demonstrates the rapidity with which novel enzyme inhibitors may be discovered and optimized through a combination of molecular modelling and combinatorial synthesis. Although in this example, compounds were selected from the VL by applying computational algorithms that picked new compounds for synthesis on the basis of geometrical similarity, other approaches to lead optimization using VLs can be applied. Genetic algorithms (GA) belong to a class of evolutionary principles of natural selection that dictate the manner in which organisms reproduce, and evolve to be fitter for their environment under the pressure of selection. In the case of drug discovery, the GA is a computational process and 'fitter' relates to greater affinity for the pharmacological target. The lead optimization process follows a series of generations of synthesis and screening, seeking more potent compounds from each successive generation of new compounds.

In an applied example of successful GA lead optimization, the Ugi four-component condensation reaction was used to generate products that were tested as inhibitors of the enzyme thrombin (Weber *et al.* 1995). The building blocks used were a comprehensive collection of 10 isonitriles, 40 aldehydes, 10 amines, and 40 carboxylic acids. The amine set was constrained to monomers that possess a basic side-chain known to confer affinity for the trypsin-like serine protease family of which thrombin is a member. The mathematical rules of combinatorial chemistry indicate that if every compound were synthesized there are a potential 160 000 possible Ugi products. Traditionally, a mix and split synthetic approach could be used to make all the possible compounds in mixtures for screening. However, the advantage of this GA strategy is that from the 160 000 compounds represented by the total size of the VL, only 20 compounds needed to be synthesized in each of the generations.

The GA strategy implemented worked as follows. Each monomer was encoded with a unique but arbitrary bit code, for example 0010 for benzylisonitrile, and each of the VL compounds would therefore be described by a bit string comprised of the appropriate monomer codes. The reaction mixture that should generate the compound **9.7** was labelled as 0010 011100 0111 01011 from the codes of the precursors: benzylisonitrile, salicylaldehyde, agmantine, and propionic acid (Fig. 9.3). The experiment commenced

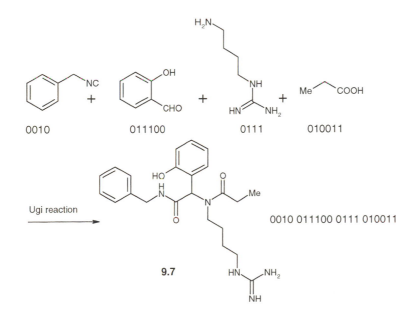

**Fig. 9.3** The synthesis of one Ugi library member and the binary codes for the monomers and reaction mixture.

More recently many more pharmaceutical companies have become involved with combinatorial chemistry, either directly or through collaborations or acquisitions of the best of the new biotechs. However, it is evident that the multifarious techniques have been appraised, and consolidation has resulted in many fewer techniques now being much more widely applied. Chemists have challenged the assumption that more is better, and instead have sought quality rather than quantity in combinatorial chemistry. The numbers of compounds synthesized in each library has decreased from the billions of peptides to smaller sets of hundreds or thousands of low molecular weight compounds. At the same time, commensurate with this decrease in library size, the mixture size has decreased to small pools of 20 or so, or in many cases the synthesis of single compounds. Making libraries and discovering active compounds in smarter rather than faster ways with a focus on design, purity, and characterization has become the new objective. Consequently, the medicinal chemists have expressed a belief in making compounds that improve the chances of drug discovery by making *more appropriate* compounds rather than just *more* compounds.

This final chapter has explored the impact of combinatorial chemistry on a range of scientific areas, although overall, it is in the biological sciences, with the search for novel pharmacological active molecules that this technology has its greatest impact. The scientific and commercial potential has instigated a thorough investigation of many different ways of generating and testing compound libraries, and the various chapters have described the numerous technologies that have been explored. We have seen how both solid phase and solution phase synthesis methods have their roles, and how newer synthetic methods are emerging that use a combination of both solution and solid phase to successfully complete a synthesis, using phase separations to maintain the quality of the final library products. The book has also described how combinatorial methods cover the range of library sizes from the very large libraries of billions of peptides, down to the tens of single compounds prepared for a specific biological target. Some libraries rely on highly sophisticated encoding techniques that allow the unambiguous identification of the active compounds, others require some synthetic commitment to find the active compounds through an iterative deconvolution of mixtures. We have also seen how combinatorial chemistry has initiated a renaissance in several related chemistry areas, especially those of solid phase synthesis and solid phase chemical analysis. Finally, we have considered how combinatorial chemistry has triggered new thinking in science, and how new techniques are emerging, that whilst being heavily influenced by combinatorial chemistry, require essentially no synthesis to find preferred library compounds.

The mark of the success of combinatorial chemistry is not only in the individual success stories of epitope studies, synthetic receptor design, or drug discoveries, but rather in its ability to capture the imagination of the innovative chemist; in its ability to generate a huge number of answers to the same scientific problems; in its ability to influence the thinking and invention of scientists in numerous different scientific fields. But more than any of this, the success of combinatorial chemistry is evident from its ability to permanently change our understanding of the nature and practise of chemistry.

# References

Huc, I. and Lehn, J-M. (1997). Virtual combinatorial libraries: dynamic generation of molecular and supramolecular diversity by self-assembly. *Proc. Natl. Acad. Sci. USA*, **94**, 2106–10.

Kaldor, S. (1997). Presented at the CHI conference: '*Exploiting Molecular Diversity*', Coronado, USA. 3rd–5th February 1997.

Kick, E.K., Roe, D.C., Skillman, A.G., Liu, G., Ewing, J.A., Sun, Y., *et al.* (1997). Structure-based design and combinatorial chemistry yield low nanomolar inhibitors of cathepsin. D. *Chem. Biol.*, **4**, 297–307.

Owns, R.A., Gesellchen, P.D., Houchins, B.J., and DiMarchi, R.D. (1991). The rapid identification of HIV protease inhibitors through the synthesis and screening of defined peptide mixtures. *Biochem. Biophys. Res. Commun.*, **181**, 402–8.

Patchett, A.A., Nargund, R.P., Tata, J.R., Chen, M-H., Barakat, K.J., Johnston, D.B.R., *et al.* (1995). Design and biological activities of L-163,191 (MK-0677): a potent, orally active growth hormone secretagogue. *Proc. Natl. Acad. Sci. USA*, **92**, 7001–5.

Rano, T.A., Timkey, T., Peterson, E.P., Rotonda, J., Nicholson, D.W., Becker, J.W., *et al.* (1997). A combinatorial approach for determining protease specificities: application to interleukin-1$\beta$ converting enzyme (ICE). *Chem. Biol.*, **4**, 149–55.

Shuker, S.B., Hajduk, P.J., Meadows, R.P., and Fesik, S.W. (1996). Discovering high-affinity ligands for proteins: SAR by NMR. *Science*, **274**, 1531–4.

Terrett, N.K., Bojanic, D., Brown, D., Bungay, P.J., Gardner, M., Gordon, D.W., *et al.* (1995). The combinatorial synthesis of a 30,752-compound library: discovery of SAR around the endothelin antagonist, FR-139,317. *Bioorg. Med. Chem. Lett.*, **5**, 917–22.

Wang, G.T., Li, S., Wideburg, N., Krafft, G.A., and Kempf, D.J. (1995). Synthetic Chemical Diversity: solid phase synthesis of libraries of $C_2$ symmetrical inhibitors of HIV protease containing diamino diol and diamino alcohol cores. *J. Med. Chem.*, **38**, 2995–3002.

Weber, L., Wallbaum, S., Broger, C., and Gubernator, K. (1995). Optimization of the biological activity of combinatorial compound libraries by a genetic algorithm. *Angew. Chem. Int. Ed. Engl.*, **34**, 2280–2.

# Index